Martin von Kurnatowski

Selection theory of dendritic growth in complex systems

Martin von Kurnatowski

Selection theory of dendritic growth in complex systems

Südwestdeutscher Verlag für Hochschulschriften

Impressum / Imprint
Bibliografische Information der Deutschen Nationalbibliothek: Die Deutsche Nationalbibliothek verzeichnet diese Publikation in der Deutschen Nationalbibliografie; detaillierte bibliografische Daten sind im Internet über http://dnb.d-nb.de abrufbar.
Alle in diesem Buch genannten Marken und Produktnamen unterliegen warenzeichen-, marken- oder patentrechtlichem Schutz bzw. sind Warenzeichen oder eingetragene Warenzeichen der jeweiligen Inhaber. Die Wiedergabe von Marken, Produktnamen, Gebrauchsnamen, Handelsnamen, Warenbezeichnungen u.s.w. in diesem Werk berechtigt auch ohne besondere Kennzeichnung nicht zu der Annahme, dass solche Namen im Sinne der Warenzeichen- und Markenschutzgesetzgebung als frei zu betrachten wären und daher von jedermann benutzt werden dürften.

Bibliographic information published by the Deutsche Nationalbibliothek: The Deutsche Nationalbibliothek lists this publication in the Deutsche Nationalbibliografie; detailed bibliographic data are available in the Internet at http://dnb.d-nb.de.
Any brand names and product names mentioned in this book are subject to trademark, brand or patent protection and are trademarks or registered trademarks of their respective holders. The use of brand names, product names, common names, trade names, product descriptions etc. even without a particular marking in this work is in no way to be construed to mean that such names may be regarded as unrestricted in respect of trademark and brand protection legislation and could thus be used by anyone.

Coverbild / Cover image: www.ingimage.com

Verlag / Publisher:
Südwestdeutscher Verlag für Hochschulschriften
ist ein Imprint der / is a trademark of
OmniScriptum GmbH & Co. KG
Heinrich-Böcking-Str. 6-8, 66121 Saarbrücken, Deutschland / Germany
Email: info@svh-verlag.de

Herstellung: siehe letzte Seite /
Printed at: see last page
ISBN: 978-3-8381-5109-0

Zugl. / Approved by: Magdeburg, Otto-von-Guericke-Universität, Diss., 2015

Copyright © 2015 OmniScriptum GmbH & Co. KG
Alle Rechte vorbehalten. / All rights reserved. Saarbrücken 2015

Acknowledgements

This book emanated from my PhD thesis at the department of theoretical physics of the Otto-von-Guericke-University Magdeburg. The corresponding referees were Prof. Dr. Klaus Kassner and Dr. Efim Brener.
I would like to thank Prof. Dr. Klaus Kassner for the excellent supervision throughout the entire time period of the work. I benefited from his expertise and learned a lot from the work in his group.
I am grateful to Jean-Marc Debierre for stimulating discussions, and I thank both, Jean-Marc Debierre as well as Rama Guerrin, for their hospitality during my stays in Marseille and, not least, for much french parlance practice.
Furthermore, I credit Kristian Lwe with parts of the C programming. His experience, suggestions and code snippets contributed to the numerical tool for selection theory with convection, which arose from this work.
The financial funding of the *German Research Foundation* (DFG) under Grant No. KA 672/10-1 is greatly appreciated.

Abstract

Dendritic patterns occur when a crystal grows in its own undercooled melt. Such a situation arises, when a germ nucleates in a liquid, the temperature of which is below its freezing point. A solid starts to grow, and the two-phase boundary takes a dendritic shape. A transport mechanism is required to remove the latent heat released at the interface. Otherwise, the vicinity of the crystal would heat up and the growth would end.

A detailed understanding of the growth procedure and the underlying physics is important, because the microscopic dendrites strongly influence the macroscopic properties of a bulk sample. It had been puzzling for a long time, how the pattern is formed, until it was found out, that anistropic surface tension provides the sought-after mechanism stabilizing a parabolic dendrite tip. From a theoretical point of view, solving the problem requires a notable effort, even if the describing model is constrained to its most basal features. If surface tension is neglected, there is a continuous family of solutions, out of which the stable growth mode has to be found at finite surface tension. Therefore, it is also called a "selection problem".

The classical solution technique is based on Green's functions. Yet, this method is restricted to the simplest models, where heat is transported solely by diffusion. That simplicity is overcome in this work. The systems under consideration here are complex in terms of the advanced heat transport properties taken into account with the purpose to investigate the effect on the selected growth parameters and to improve the understanding of the method used here as well as testing its scope of applicability. The models are treated by means of asymptotic decomposition. The method is predominantly analytical, apart from the last step, for which a C program was written.

A considerable part of the work is concerned with convective models. The inclusion of flows in the liquid phase drastically complicates the problem, since the field equations are rendered nonlinear. Analytical solutions of the dendritic growth problem with convection are derived in this work. A forced flow velocity is introduced as additional parameter in order to predict its capability of controling the growth. It is found, that convection changes the scaling laws in the system, but its influence on the selected eigenvalue in the considered range for the flow velocity is marginal.

Moreover, nonlinear diffusion, thermal resistance and anisotropic diffusion are considered. In the latter case, a rigorous analytical solution is presented and it is found, that the anisotropy of diffusion cannot take the stabilizing role of surface tension.

Contents

1. **Basic concepts and approaches to dendritic growth**   **9**
   1.1. Physics of solidification described by a macroscopic continuum model 9
   1.2. Singular perturbation theory in the symmetric model . 21
        1.2.1. Ivantsov's needle crystals 21
        1.2.2. Growth mode selection using Green's functions 25
   1.3. Extended experimental and theoretical study of the selected pattern 32
        1.3.1. Features of free dendritic growth 32
        1.3.2. Previous analytical treatment of dendritic growth in convective systems 35
        1.3.3. Uncertainty about crystalline anisotropy strength measurements in relevant materials . . 38

2. **Advanced approach for dendritic growth in nonlinear systems**   **41**
   2.1. Conformal parabolic coordinates and non-dimensionalization 41
   2.2. Demonstration of the analytical part of the method by application to the symmetric model 45
   2.3. Numerical treatment of the local equation 52

3. Convective problems 61

3.1. Potential flow 61

    3.1.1. Ivantsov solution for dendritic growth in a potential flow 63

    3.1.2. Continuation to the complex plane and asymptotic decomposition (potential flow) 66

    3.1.3. WKB analysis of the linearized equation far from the singularity (potential flow) 68

    3.1.4. Transformation to a small disk around the singularity and asymptotic matching to the WKB solution (potential flow) 72

    3.1.5. Numerical results and dependencies of the observable quantities and the selected growth mode on the potential flow 75

3.2. Oseen flow 79

    3.2.1. Ivantsov solution for dendritic growth in an Oseen flow 80

    3.2.2. Continuation to the complex plane and asymptotic decomposition (Oseen flow) 86

    3.2.3. WKB-analysis of the linearized equation far from the singularity (Oseen flow) 89

    3.2.4. Transformation to a small disk around the singularity and asymptotic matching to the WKB solution (Oseen flow) 92

    3.2.5. Scaling laws in the large flow Péclet number limit 93

    3.2.6. Numerical results and dependencies of the observables quantities and the selected growth mode on the Oseen flow 99

4. Extended heat transport properties   **111**

4.1. Nonlinear diffusion 111

     4.1.1. Ivantsov solution for dendritic growth in a system with temperature-dependent thermal diffusivity 111

     4.1.2. Continuation to the complex plane and asymptotic decomposition (nonlinear Diffusion) 115

4.2. Thermal resistance at the two-phase boundary 119

     4.2.1. Ivantsov solution for dendritic growth with thermal interface resistance 121

     4.2.2. Shape equation in parabolic and cartesian coordinates for dendritic growth with finite thermal resistance 125

4.3. Anisotropic diffusion 129

     4.3.1. Rescaled system and Ivantsov solution for dendritic growth with anisotropic diffusion 131

     4.3.2. Derivation of the shape equation and its WKB solution 135

     4.3.3. Derivation of the local equation and its numerical solution for dendritic growth mode selection with anisotropic diffusion 136

4.4. Arbitrary growth Péclet numbers and asymptotic decomposition 144

     4.4.1. Transformation of the temperature field 145

     4.4.2. Continuation to the complex plane and asymptotic decomposition (arbitrary Péclet numbers) 146

4.5. Kinetic effects 152

     4.5.1. Limit of isotropic kinetic effects and anisotropic surface tension 155

     4.5.2. Limit of isotropic surface tension and anisotropic kinetic effects 156

5. Conclusion   **159**

A. Auxiliary calculations   **163**
   A.1. Potential flow 163
       A.1.1. Setup of the model equations 163
       A.1.2. Asymptotic decomposition 165
       A.1.3. WKB solution 169
       A.1.4. Local equation 173
   A.2. Oseen flow 177
       A.2.1. Ivantsov solution 177
       A.2.2. Asymptotic decomposition 181
       A.2.3. WKB solution 189
       A.2.4. Local equation 198
   A.3. Thermal resistance in cartesian coordinates 205
       A.3.1. Interface normal, curvature and anisotropy function in cartesian coordinates 205
       A.3.2. Derivation of the cartesian shape equation (4.43) 207
   A.4. Anisotropic diffusion 211
       A.4.1. Expansion of the boundary conditions 211
       A.4.2. WKB solution 214
       A.4.3. Stretching transformation of the shape equation 215

B. C-code demo   **217**
   B.1. Oseen flow subroutines parallel to the imaginary axis . 218
   B.2. Oseen flow subroutines on the real axis 227

C. Material parameters   **233**

Bibliography   **235**

1. Basic concepts and approaches to dendritic growth

1.1. Physics of solidification described by a macroscopic continuum model

Solidification is a phenomenon which is known to everyone just from common experience, for example regarding a freezing lake. From a technical point of view, it is important in many production processes. For instance for the casting of metals or for welding, fundamental understanding of solidification is needed to control these processes in a satisfactory manner. The microscopic material properties developed during solidification heavily affect the macroscopic properties of the solid bulk. In fact, solidification is included in the manufacturing of most products at a certain level. Even semiconductors for microelectronics are still obtained from the liquid phase if bulk material is needed. When solidification is the final production stage, its effect is even more important [77].

Physically, solidification is a first order phase transition. It happens in a liquid at melting temperature or at an even lower temperature (undercooled liquid), when a solid nucleus has exceeded a critical size. The solid begins to grow, and a two-phase boundary is formed. Three different aspects influence the growth process: diffusion, interface kinetics and capillarity. These are discussed in the following in that very order. The total differential of the *Gibbs free energy* per particle

1. Basic concepts and approaches to dendritic growth

is $dg = vdp - sdT$, and its first derivatives behave discontinuously, i.e. the volume per particle v and the entropy per particle s show a jump behaviour at the melting temperature. The discontinuity of the entropy per particle implies the release of the latent heat

$$l = T_M(s^l - s^s) \qquad (1.1)$$

per particle. Here the superscripts l and s point out the quantities in the liquid and the solid respectively, and T_M is the melting temperature of the equilibrium interface. The latent heat l can be understood from the microscopic properties of the material. When the spatial alignment of the particles exhibits a far field order, then the solid is a crystal. Thus, one also speaks of crystal growth. The particles are usually packed denser than in the liquid phase. The reduction of the mean particle distance is accompanied by a decrease in the interparticle potential energy. It happens isothermally at $T = T_M$. But after the phase transition, the temperature at the liquid-solid interface increases due to the released latent heat, and growth will stop unless the heat is removed by some transport mechanism. One such mechanism is heat diffusion.

To prevent the interface temperature from exceeding T_M, one has to cool the system externally. Indeed there is also crystal nucleation in liquids with $T > T_M$, but these nuclei are unstable. One distinguishes between two modes of growth: columnar and equiaxed growth. When a substance solidifies in a vessel, solid nuclei are much more likely to form close to the walls, because there the temperature is lower due to the external cooling and there are more inhomogeneities. These crystals can grow in any direction, and they form the outer equiaxed zone. However, growth is faster parallel or antiparallel to the favoured heat flux direction, which is perpendicular to the walls. Crystals growing inwards (away from the walls) outgrow the others, and they align in a zone of distinct columnar blocks [77]. In columnar growth, the solid is cooled directly at its contact area with the vessel walls. The heat

1.1. Physics of solidification described by a macroscopic continuum model

flux is antiparallel to the growth direction, and the heat is transported through the solid. This growth mode is favoured, if the absolute values of the local temperature gradients are rather large. Beyond a certain stage, small branches, that detach from the columnar crystals, can grow independently. They take an equiaxed shape and form the inner equiaxed zone. In equiaxed growth, the heat flux is largely parallel to the growth direction [77]. A negative thermal gradient in the liquid enables continuous growth. Equiaxed growth is the favoured growth mode in systems with high latent heat production rate. The onset of equiaxed growth is more likely, if fast flows in the melt are present, and electromagnetic stirring is often used to promote the transition, because equiaxed growth yields superior material properties of the solid sample for the majority of applications [77].

A different situation arises in the case of directional solidification, where the constant temperature gradient and the growth velocity are forced experimentally by pulling a crucible through a specimen, leading to different growth regimes [95]. Often this is technically favourable. In contrast to that, in free growth both the temperature profile and the growth velocity are selected by the system and have to be calculated in a theoretical framework. The special selection mechanism will be discussed later. In multi-component systems, such as metal alloys, the incorporation of particles into the solid happens on a fast time scale and the growth is governed by material diffusion. In this case, a chemical model instead of a thermal model is needed. But this work is concerned almost only with thermal models. Anyway, in both cases the model equations have the same mathematical structure [69].

Interface kinetics can also influence the growth procedure. In crystal growth, the probability of adsorption of a particle to the solid surface increases with the number of nearest lattice neighbours at the free surface site. Kinetic effects are relevant, if their timescale is slow and less latent heat is produced [77]. They affect the lattice planes, which are observed as the solid surface during facetted growth. It can be investigated using microscopic models, such as diffusion-limited aggregation (DLA). Thermal diffusion plays a more important role for the growth, if the time scale of kinetic effects is fast. In this case, considerably more particles get attached to the solid at higher diffusion rates in the liquid. In this work, the growth is always non-facetted, the liquid-solid interface is microscopically rough and macroscopically smooth. Hence, the relevance of interface kinetics is assumed to be only marginal for most of the calculations done here. Yet, kinetic effects on the considered systems have been analyzed for completeness (see subsection 4.5).

The shape of the two-phase boundary is basically determined by capillarity if interface kinetics is negligible. If the melt temperature is lower than the interface temperature, then the thermal gradient is negative and planar solidification fronts with constant growth velocity are prohibited by heat conservation [82]. They could only persist, if the released latent heat was completely absorbed by the solid, which is not the case in experiments. They are morphologically unstable due to the *Mullins-Sekerka instability* [90]. As soon as material is attached to the solid and a small protuberance evolves somewhere on the interface, the isotherms are deformed and the temperature in the liquid has to decay by the same amount in a slightly smaller range. The thermal gradient becomes steepest at the tip of the protuberance. Hence, heat is removed most efficiently there, and it is the favoured location for the attachment of additional material. The tip grows faster than the remainder of the two-phase boundary, and a dendrite evolves. The dendritic pattern is then stabilized by anisotropic capillary effects. In turn, this explanation of the instability shows, why planar solidification fronts may occur in the columnar growth mode, where the thermal gradient is positive. The term *"dendrite"* comes from the greek word $\delta \acute{\varepsilon} \nu \delta \rho o \nu$ (*dendron*), which means *"tree"*. It is chosen, because the microstructure often has a *tree-like* form. This

1.1. Physics of solidification described by a macroscopic continuum model

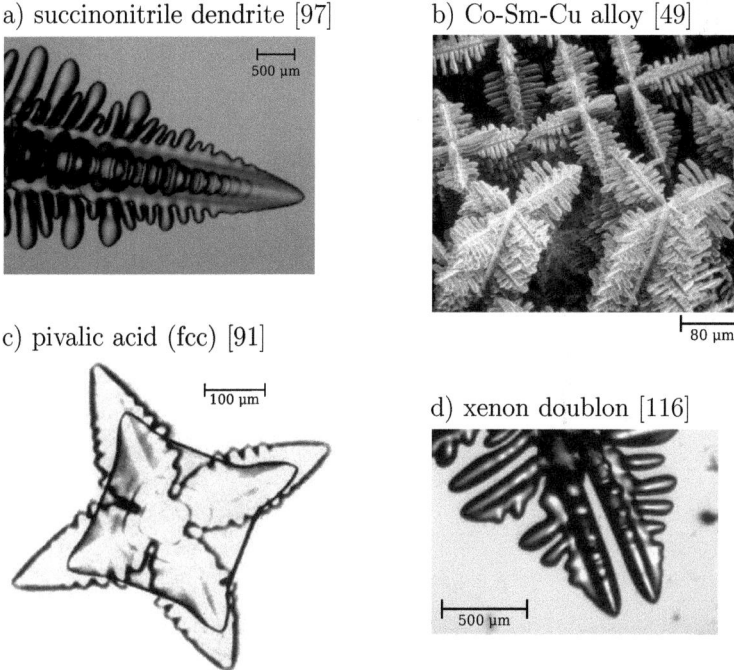

a) succinonitrile dendrite [97]   b) Co-Sm-Cu alloy [49]

c) pivalic acid (fcc) [91]

d) xenon doublon [116]

Figure 1.1.: Experimental images of dendritic microstructures in different substances: a) succinonitrile dendrite [97], b) multiple Co-Sm-Cu alloy crystals [49], c) pivalic acid crystal with a shape resulting from the underlying fourfold fcc-lattice symmetry [91], d) doublon structure [116]

process of pattern formation in diffusion-limited crystal growth has been an object of research for years [82]. A beautiful review can be found in [11]. For pattern formation to occur, a non-equilibrium situation is often necessary. This is established by the external cooling. The undercooling is an external control parameter. Without external cooling, the latent heat increases the temperature in the vicinity of

1. Basic concepts and approaches to dendritic growth

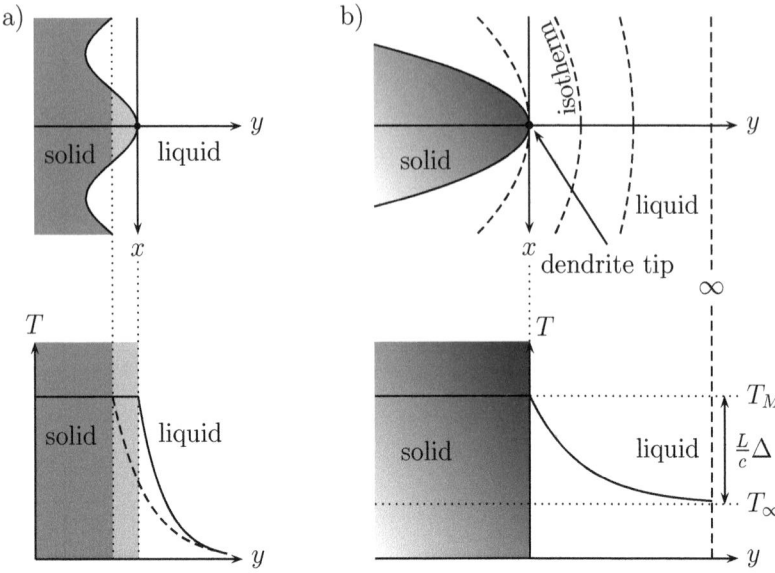

Figure 1.2.: a) A planar solidification front is destroyed by an instability (top). The protuberance is marked by a lighter color than the rest of the solid. The corresponding temperature profile along $x = 0$ (bottom picture) shows the negative thermal gradient in the liquid. It is steeper at the tip compared to the planar front (dashed graph), because the temperature has to decay by the same amount in a slightly smaller range. This causes the dendrite to grow. The solid can be considered essentially isothermal. b) The growing dendrite in real space (top). The isotherms are sketched by black dashed lines. Corresponding temperature profile (bottom picture) with an illustration of the dimensionless undercooling Δ defined in equation (1.18).

1.1. Physics of solidification described by a macroscopic continuum model

the two-phase boundary up to T_M and the crystal ceases to grow.

Figure 1.1 exhibits four examples of dendritic microstructures in different substances. The pictures are taken from the cited experimental works. Image 1.1b shows multiple dendritic crystals on a larger length scale. From this, one can get a notion of how the amount of tilt and interlocking of the dendrites can affect the robustness of a macroscopic sample. Though, the mere size of the patterns is more essential for the macroscopic properties. Image 1.1c reveals the influence of the underlying atomic lattice. The fourfold symmetry of the face-centered cubic *Bravais lattice* is reflected by the equilibrium crystal shape. Mathematically, this is realized by a well-chosen anisotropy function of interface surface tension. Figure 1.1d exhibits an example of a doublon structure, which can be observed under certain system conditions. They are explained a bit more in one of the upcoming paragraphs. Figure 1.2a illustrates the instability driven by different temperature gradients, and figure 1.2b exhibits the growing dendrite (actually only a needle crystal without sidebranches) in real space with a sketch of the isotherms.

After long times, the crystal ceases to grow and takes an equilibrium shape, which can be determined by a graphical technique called Wulff construction [65, 109]. It is based on a minimization of the free surface energy, the orientational dependence of which is a consequence of the underlying lattice structure. In the systems here considered, the capillary effects tend to maximize the length scale of the pattern, whereas diffusion tends to minimize the length scale of the pattern. The result is a compromise between the two competing processes. The favoured direction for dendritic growth is the axis of maximal surface tension or minimal surface stiffness. Furthermore, the nonlinearity needed to describe pattern formation mathematically arises in the problem's Green's function. But the shape also appears nonlinearly in a capillar interface condition, which shall be derived in the following. Consider

1. Basic concepts and approaches to dendritic growth

the change rate of the *Gibbs free energy* G at the interface:

$$\frac{dG}{dt} = \frac{\partial G^s}{\partial V^s}\frac{dV^s}{dt} + \frac{\partial G^l}{\partial V^l}\frac{dV^l}{dt} + \frac{\partial G^s}{\partial A^{sl}}\frac{dA^{sl}}{dt}. \qquad (1.2)$$

Here, the superscripts s and l indicate the solid and the liquid domain respectively. A^{sl} is the area of the two-phase boundary. The total differential of G is $dG^{s,l} = V^{s,l} dp - S^{s,l} dT$. Assuming the pressure to be constant ($dp = 0$), one finds

$$\frac{\partial G^{s,l}}{\partial V^{s,l}} = -\frac{S^{s,l}}{\Omega^{s,l}} \int_{T_M}^{T} dT' = -\frac{S^{s,l}}{\Omega^{s,l}} (T - T_M). \qquad (1.3)$$

Here, $\Omega^{s,l}$ denotes the volume per mole in the respective phase. This differential expression may be considered accurate, if the solid and the liquid phases are incompressible [53]. Moreover, the surface tension γ is defined by

$$\frac{\partial G^{s,l}}{\partial A^{sl}} = \gamma. \qquad (1.4)$$

γ is positive for any well-defined structure. Mass conservation is written as

$$\frac{1}{\Omega^s}\frac{dV^s}{dt} = -\frac{1}{\Omega^l}\frac{dV^l}{dt}. \qquad (1.5)$$

In a local equilibrium situation, one has $\dot{G} = 0$ at the interface. (1.3), (1.4) and (1.5) are put together into (1.2):

$$0 = \left[\frac{S^l - S^s}{\Omega^s}(T - T_M) + \gamma \frac{dA^{sl}}{dV^s}\right]\dot{V}^s. \qquad (1.6)$$

Insofar as the growth rate $\dot{V}^s$ of the solid is arbitrary, the term in the square bracket has to vanish. $\frac{dA^{sl}}{dV^s}$ can be identified with the curvature κ. Introducing the latent heat per unit volume $L = T_M(S^l - S^s)/\Omega^s$, one ends up with

$$T = T_M - \frac{T_M}{L}\gamma\kappa. \qquad (1.7)$$

1.1. Physics of solidification described by a macroscopic continuum model

Equation (1.7) applies at the liquid-solid interface [53]. It shows that due to capillarity, the equilibrium temperature of the interface depends on its shape (*Gibbs-Thomson effect*).[1] The selected shape and growth velocity, which are experimentally measurable, correspond to a structurally stable solution of a model. It turns out that the underlying selection mechanism is non-trivial.

After all these considerations, a model describing free solidification can be written down. We examine the two-dimensional problem of a crystal growing in its undercooled melt. There may be convective flows in the liquid phase. Furthermore, we make the assumption, that the thermal diffusivity D is the same in the liquid and in the solid. The same is assumed for the mass density ϱ_m and the volume specific heat capacity c at constant pressure. The governing equations for this problem are the temperature diffusion equation in the solid and the temperature diffusion-advection equation in the liquid:

$$\frac{\partial T^s}{\partial t} = \vec{\nabla} \cdot \left(D \cdot \vec{\nabla} T^s \right) \qquad \text{in the solid,} \qquad (1.8a)$$

$$\frac{\partial T^l}{\partial t} + \left(\vec{w} \cdot \vec{\nabla} \right) T^l = \vec{\nabla} \cdot \left(D \cdot \vec{\nabla} T^l \right) \qquad \text{in the liquid.} \qquad (1.8b)$$

These equations are phenomenological bulk equations with the temperature T being the relevant field quantity. They are a continuum representation of energy conservation. $\vec{w}$ is the flow velocity. In general D is a tensor and it can also depend on the temperature T (nonlinear diffusion) [76]. The advection term brings a nonlinear coupling of the field equations into the system. Note that the inclusion of convection adds tremendous complexity to the problem. This effect has been neglected in most analytical works on dendritic growth. The boundary conditions, valid at the interface, are

$$T^s = T^l \qquad \text{continuity} \qquad (1.9a)$$

---

[1] The instability mechanism destroying planar solidification fronts would also work on an isothermal interface.

1. Basic concepts and approaches to dendritic growth

$$T^s = T_M - \frac{L}{c}d_0\kappa a(\theta) - \frac{L}{c}\bar{\beta}(\theta)\vec{V}\cdot\vec{n} \quad \text{Gibbs-Thomson} \quad (1.9b)$$

$$\frac{L}{c}\vec{V}\cdot\vec{n} = D\cdot\left(\vec{\nabla}T^s - \vec{\nabla}T^l\right)\cdot\vec{n} \quad \text{Stefan condition} \quad (1.9c)$$

with

$\vec{V}$ the growth velocity,

d_0 the mean capillarity length,

$\bar{\beta}(\theta)$ the anisotropy function of kinetic effects,

$a(\theta)$ the anisotropy function of capillary effects with θ being the angle between the normal to the interface and a fixed direction,

$\vec{n}$ the normal vector to the interface, pointing into the liquid.

They apply at the moving interface. Far away from the interface, homogeneous *Dirichlet boundary conditions* are imposed. I.e. the temperature in the liquid must take a constant value T_∞, which can be controlled in experiments by external cooling, and the temperature in the solid must approach T_M:

$$T^l \to T_\infty \quad (1.10a)$$
$$T^s \to T_M \quad (1.10b)$$

for $|\vec{r}| \to \infty$ in either the liquid or the solid. Condition (1.9a) is the temperature *continuity condition* at the two-phase boundary. Condition (1.9b) was obtained from (1.7). In general, the surface tension $\gamma(\theta)$ is orientation-dependent, and then the surface stiffness $\gamma(\theta)+\gamma''(\theta)$ is the relevant prefactor of the curvature. This is included in the product $d_0 a(\theta)$. The mean capillary length $d_0 = T_M\gamma_0 c/L^2$ is proportional to the mean surface tension γ_0 averaged over all angles θ. Equation (1.7) without the last term is called *Gibbs-Thomson condition* describing capillary effects under the assumption of local equilibrium. It has been extended by a term representing kinetic effects of

1.1. Physics of solidification described by a macroscopic continuum model

atom transfer at the liquid-solid interface, and it therefore describes a weakly non-equilibrium situation. The *Stefan condition* (1.9c) represents local enthalpy conservation at the phase transition, and it supplements (1.8a) and (1.8b) to ensure energy conservation everywhere. Different mass densities in the two phases would significantly increase the complexity of the problem. Material would get sucked towards the interface, and the symmetry in the Stefan condition gets broken. The flow velocity $\vec{w}$ is determined by the *Navier-Stokes equations* for incompressible fluids:

$$\vec{\nabla} \cdot \vec{w} = 0 \qquad \text{incompressibility,} \qquad (1.11)$$

$$\frac{\partial \vec{w}}{\partial t} + \left(\vec{w} \cdot \vec{\nabla}\right) \vec{w} = -\frac{1}{\varrho_m}\vec{\nabla} p + \nu \Delta \vec{w} \qquad \text{Navier-Stokes.} \qquad (1.12)$$

with the kinematic viscosity ν. The Navier-Stokes equation (1.12) is the momentum equation (*Newton's second law*) arranged for viscous Newtonian fluids. It can also be derived from classical statistical physics, where a microscopic description turns out to be futile and the assumption that the mean in the phase space equals the mean in time, leads to macroscopic differential equations. The boundary conditions for $\vec{w}$ are

$$\vec{w} \cdot \vec{n} = 0 \qquad \text{mass conservation} \qquad (1.13a)$$
$$\vec{w} \cdot \vec{t} = 0 \qquad \text{no-slip} \qquad (1.13b)$$

at the interface and

$$\vec{w} = -U\vec{e}_y \qquad (1.14)$$

far ahead of the interface ($|\vec{r}| \to \infty$ in the liquid). $\vec{n}$ and $\vec{t}$ are the normal and tangential unit vectors to the interface respectively. The boundary condition (1.13a) is the *mass conservation* condition. Here the mass densities of the solid and the liquid are assumed to be equal. Condition (1.13b) holds due to friction, and is sometimes called the *no-slip condition*. In the far field boundary condition (1.14), there is a constant forced flow velocity $-U\vec{e}_y$. This is an important external parameter characterizing the flow. The forced flow is supposed to be anti-parallel to the growth direction, and the interface is flooded with undercooled liquid. It may be considered a simplified modelling of buoyancy or natural convection. But it can also be forced artificially in real systems in order to control the growth process.

Field equations (1.8a),(1.8b) together with boundary conditions (1.9a), (1.9b), (1.9c), (1.10a) and equations (1.11), (1.12) together with boundary conditions (1.13a), (1.13b), (1.14) represent a macroscopic continuum description of free solidification. The model is a nonlinear free boundary problem. Determining the position and the shape of the interface is part of the problem. That is why the information contained in the additional condition (1.9c) is needed, although (1.9a)-(1.9b) together with (1.10a)-(1.10b) already specify values for T on the whole boundaries of the domains. There is also a one-sided model [82], which is focused only on the liquid domain whereas heat diffusion in the solid is ignored. This is especially a reasonable simplification in the chemical model, because impurity diffusion in the solid is so much smaller than in the liquid, that it can be neglected altogether [82]. But the model introduced above is already simple enough for most parts of this work and yet quite feasible, because it can be used for a variety of problems. Concerning the title of this work, the systems described by the model can be "complex" in the sense that the heat transport may be rather sophisticated. Convection, nonlinear diffusion or anisotropic diffusion may be taken into account and these aspects remarkably complicate the problem. However, the equations can also simplify when being adapted to a particular problem.

One of such simplifications is the following: In all the upcoming chapters we are going to look for solutions with a shape-preserving solidification front growing with a steady-state velocity V. Such quasi-stationary solutions are implied by experimental observations [39]. Thus, the equations can be rewritten in a moving frame. The only

field equation that changes is equation (1.8a). The flow equations
(1.11)-(1.12) as well as the diffusion-advection equation (1.8b) are invariant under *Galilean transformation*. (For (1.8b), this applies unless
flow is neglected.) The results of this work are presented in chapters
3 and 4. The effect of convection on the growth of dendritic patterns
is analyzed in detail in chapter 3 of this work. Two different approximations for the flow velocity field are used: A potential flow and the
more realistic but more complex Oseen flow. An analytical solution is
derived, which is conjectured to be the most accurate one so far. The
boundary conditions are evaluated numerically, providing explicitly
the dependence of the growth parameters on the flow. In chapter 4,
extended aspects of dendritic growth are considered (nonlinear diffusion, thermal resistance, anisotropic diffusion, arbitrary growth Péclet
number, kinetic effects) in order to investigate the scope of the used
method. But before, the used analytical and numerical tools are introduced in chapter 2, allowing for a better understanding and a more
compact presentation in the subsequent chapters.

1.2. Singular perturbation theory in the symmetric model

The most successful works on dendritic growth employed the *symmetric model*. It is the convection-free form of the general model introduced in the preceding section 1.1. In addition to that, the diffusion
coefficient D is assumed to be the same in the solid and in the liquid.

1.2.1. Ivantsov's needle crystals

In experiments, a nearly parabolic shape of the dendritic interface
pattern is observed at the tip, i.e. ahead of the sidebranches [101].
In fact, a parabolic solution to the symmetric model can be found, if

1. Basic concepts and approaches to dendritic growth

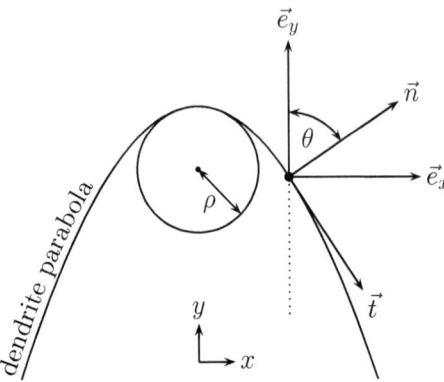

Figure 1.3.: The growing parabolic dendrite, illustration of important vectorial quantities

surface tension is neglected. The Gibbs-Thomson condition without kinetic effects has to be simplified:

$$T^s = T_M. \tag{1.15}$$

Hence, the neglect of surface tension results in an isothermal interface, because equation (1.15) is obtained from (1.9b) by setting $d_0 = 0$ and the capillary length d_0 is proportional to the average surface tension γ_0. The interface takes the form

$$y_s(x) = \frac{\rho}{2}\left(1 - \frac{x^2}{\rho^2}\right) \tag{1.16}$$

with ρ being the radius of curvature at the growing dendrite's tip (see fig. 1.3). The whole solid domain is isothermal at $T = T_M$. The solution is related to the far field temperature by the equation

$$\Delta = \sqrt{\frac{\pi P_c}{2}} e^{\frac{P_c}{2}} \operatorname{erfc}\left(\sqrt{\frac{P_c}{2}}\right) \tag{1.17}$$

1.2. Singular perturbation theory in the symmetric model

with the dimensionless undercooling

$$\Delta = \frac{T_M - T_\infty}{L/c} \qquad (1.18)$$

and the *growth Péclet number*

$$P_c = \frac{\rho V}{D}. \qquad (1.19)$$

The *complementary error function*

$$\text{erfc}(x) = \frac{2}{\sqrt{\pi}} \int_x^\infty e^{-t^2} dt \qquad (1.20)$$

is used in (1.17). Equations of the type of (1.17) are called *Ivantsov condition* within this work, because they result from inserting the solution for the case of zero surface tension (Ivantsov solution) into the far field boundary condition. (1.17) determines the product of the growth velocity V and the tip curvature radius ρ at given undercooling Δ, but these parameters cannot be calculated separately from the equation. This family of solutions had first been found by Ivantsov [66]. Unfortunately, all of these solutions are unstable. In addition to that, a whole family of solutions seems to be rather unphysical because in experiments with a certain substance, the same growth parameters V and ρ are always observed for the same undercooling. Equation (1.17) can be simplified to $\Delta \approx \sqrt{\pi P_c/2}$ in the case of small undercooling ($\Delta \ll 1$).

Most of the works on dendritic growth are based on perturbation expansions about Ivantsov's solution as the solution of the unperturbed problem. It is a selection problem, because Ivantsov's continuous solution spectrum is converted into a discrete one, out of which only one solution is stable. This is going to be the experimentally observed solution. It is shown in detail in the next subsection. A regular perturbation expansion about an unstable solution cannot lead to any stable solutions. Instead, singular perturbation theory has to be used. In general, a system is said to be singularly perturbed, if the structure of its describing equations undergoes a significant change when a small parameter is set zero. Such structural changes can be a decrease in the order of the equation or a change of type (for instance elliptic → parabolic or PDE → ODE). In this case, the small parameter is the stability parameter σ (defined in the next subsection) proportional to d_0, and when it is set zero, the Gibbs-Thomson condition loses two derivatives and the structure of the solution space is drastically altered. When allowing for a finite value of σ, surface tension is included in the problem and it becomes technically much more challenging. Once a solution of the perturbed problem is found, it does not become equivalent to the Ivantsov solution for $\sigma \to 0$ because this limit is singular.

There used to be the hypothesis that the system selects its operating state at the point that separates stable and unstable regions of the solution space (marginal stability hypothesis [81, 83]). In this respect, the tip curvature radius ρ scales as a function of the capillary length d_0. It turned out to be wrong, although it is often in agreement with experimental data [51, 64]. One reason why the hypothesis fails is that it is based on a stability analysis of a solution with *isotropic* surface tension, but it was later shown, that no solution exists to this particular problem [24]. The anisotropic capillary effects arising from surface tension have to be taken into account to gain access to the real selection mechanism. The concept is also referred to as *"microscopic solvability theory"*: Capillarity provides a mathematical and physical solvability criterion or selection criterion. The anisotropy of surface tension gives rise to multiple coldest points on the surface of the growing dendrite [13, 15]. This stabilizes the pattern against tip splitting.

1.2. Singular perturbation theory in the symmetric model

1.2.2. Growth mode selection using Green's functions

In the first order of a singular perturbation expansion about Ivantsov's solution, capillary effects must not be neglected. The Gibbs-Thomson condition (1.9b) contains a term, which is linear in the capillary length d_0. It is a small term, since d_0 is usually of the order of several nanometres. But this additional length scale in the system is exactly what is needed to break the degeneracy of the Ivantsov solution spectrum. Let $y_s(x,t)$ again be the spatial function describing the position of the moving interface at time t. Unfortunately, this cannot be a plainly parabolic expression anymore. Barber et al. [13] used *Green's functions* to solve the diffusion equation (1.8a) and inserted it into the Gibbs-Thomson condition (1.9b), again neglecting kinetic effects:

$$\Delta - d_0 \kappa a(\theta) = \frac{P_c}{4\pi} \int_0^\infty \int_{-\infty}^\infty \frac{\dot{y}_s(x, t-\tau)}{t}$$
$$\times \exp\left[-\frac{P_c}{4\tau}\left((x-x')^2 + (y(x,t) - y(x', t-\tau))^2\right)\right] d\tau dx'.$$
(1.21)

The same approach was taken by Ben Amar and Pomeau [19] as well as by Caroli et al. [29]. This boundary integral equation[2] holds in the rest frame of the bulk. The curvature κ is a nonlinear function of $y_s(x,t)$. Actually, (1.21) is a retarded equation [69]: At $\tau = t - t' < 0$ there is no contribution to the integral because of the causality principle. The solution $y_s(x,t)$ cannot be influenced by the solution $y_s(x, t-\tau)$ at $\tau < 0$, because $t-\tau$ designates a later point in time for $\tau < 0$. The steady-state shape of the solidification front is close to the Ivantsov parabola:

$$y_s(x,t) = t - \frac{x^2}{2} + \zeta(x).$$
(1.22)

---
[2]Note, that the minus sign on the left hand side of (1.21) is a plus sign in the orginal work of Brener and Mel'nikov [24], because a different algebraic sign of the curvature will be chosen in this work. This is also accounted for in equation (1.23).

1. Basic concepts and approaches to dendritic growth

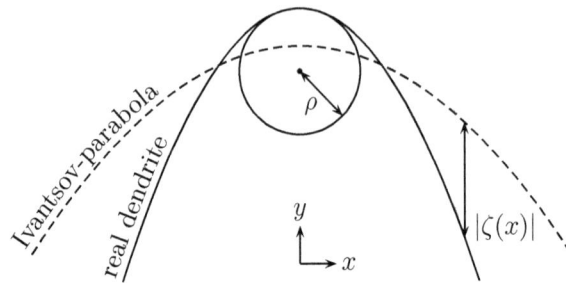

Figure 1.4.: Expanding about the Ivantsov-parabola, the real dendrite shape is corrected by a function $\zeta(x)$ and is non-parabolic

Here, lengths are measured in units of the tip curvature radius ρ. The shape of the solification front is slightly altered and non-parabolic (see fig. 1.4). $\zeta(x)$ is the shape correction function to be determined. Actually, a regular perturbation expansion can be used to calculate $\zeta(x)$ [24] with sufficient accuracy. But a singular perturbation expansion is necessary to determine the growth velocity V. Because of the smallness of the anisotropy strength of the capillary effects, $\zeta(x)$ can be assumed to be small. But its derivatives need not necessarily be small for all values of their arguments. The time integration is done using the *modified Bessel functions* K_0 and $K_1 = -K_0'$ (*Macdonald functions*) of the zero and first order respectively and $\tilde{R} = \left((x-x')^2 + \frac{1}{4}(x^2-x'^2)^2\right)^{1/2}$ [7]:

$$\sigma\kappa = \frac{P_c}{a(\theta)} \int_{-\infty}^{\infty} [\zeta(x) - \zeta(x')] \frac{e^{\frac{P_c}{2}(x^2-x'^2)}}{2\pi} \times \left[K_0\left(P_c\tilde{R}\right) - \frac{x^2-x'^2}{2\tilde{R}} K_1\left(P_c\tilde{R}\right)\right] dx'. \quad (1.23)$$

The undercooling Δ has been replaced using the Ivantsov condition (1.17). For not too large growth velocities, the diffusion field in the

1.2. Singular perturbation theory in the symmetric model

solid follows the interface adiabatically, and its profile is constant on the time scale of diffusion. The sought-after solution of equation (1.23) is stationary in a moving frame of reference attached to the dendrite growing at constant velocity V. The stability parameter

$$\sigma = \frac{2d_0}{\rho P_c} \qquad (1.24)$$

is the key to the solution of the selection problem. It is the only remaining parameter group. At small growth Péclet number P_c, σ can be considered independent of the undercooling Δ. Martine Ben-Amar investigated the problem at arbitrary undercooling and found that there is indeed a dependence of the stability parameter on Δ at large P_c [4]. Anyway, finding σ as an eigenvalue from the boundary integral equation (1.23) will select a solution because together with the Ivantsov condition (1.17), it provides complete parameter separation for the steady-state growth velocity V and the tip curvature radius ρ.

Equation (1.23) is sometimes also called the *Nash-Glicksman equation* [117]. Kessler and Levine solved it by direct numerical integration [72], and subsequently they performed a numerical stability analysis [71]. They found a stable solution of the problem with anisotropic surface tension at low undercooling using the selection criterion of a smooth tip. In the case of isotropic surface tension, the cusp magnitude at the tip is $\propto e^{-1/\sqrt{\sigma}}$, and thus it never vanishes and no solution exists. Meiron solved a quasistationary approximation of the time-dependent boundary integral equation numerically [87], and Saito et al. calculated a numerical solution of the one-sided model [103].

Others [24, 82] continued analytically: The growth Péclet number P_c is also assumed to be small ($P_c \ll 1$). This is the relevant case in most experiments. Thus, the modified Bessel functions can be expanded for

1. Basic concepts and approaches to dendritic growth

small arguments and one finds

$$\sigma a(\theta) \frac{\zeta''(x) - 1}{[1 + (\zeta'(x) - x)^2]^{3/2}} = \frac{1}{2\pi} \int_{-\infty}^{\infty} \frac{[\zeta(x') - \zeta(x)](x + x')}{\left[1 + \frac{1}{4}(x + x')^2\right](x - x')} \, dx'. \tag{1.25}$$

The explicit form of the curvature κ in cartesian coordinates has been inserted. This equation appears in the works of Langer and Barbieri [82, 14] as well as in the work of Brener and Mel'nikov [24]. In these works, different analytical procedures and physical solvability criteria are used. Langer introduces a new dependent function

$$Z(x) = \frac{\zeta(x)}{(1 + x^2)^{3/4}} \tag{1.26}$$

and obtains

$$(\mathfrak{D}_2 + \mathfrak{J}_2) * Z(x) = \frac{\sigma}{(1 + x^2)^{3/4}} \tag{1.27}$$

where the self-adjoint differential operator $\mathfrak{D}_2$ and the integral operator $\mathfrak{J}_2$ are defined by

$$\mathfrak{D}_2 = \sigma \frac{d^2}{dx^2} + \frac{\sqrt{1 + x^2}}{a(\theta)} + \mathcal{O}(\sigma). \tag{1.28a}$$

$$\mathfrak{J}_2 * Z(x) = \frac{(1 + x^2)^{3/4}}{2\pi a(\theta)} \int_{-\infty}^{\infty} \frac{\mathcal{P}}{(x - x')} \frac{(1 + x'^2)^{3/4}(x + x')}{\left[1 + \frac{1}{4}(x + x')^2\right]} Z(x') \, dx'. \tag{1.28b}$$

$\mathcal{P}$ denotes the *Cauchy principal value*, which is necessary because $\zeta(x)$ and $\zeta(x')$ are evaluated separately inside the integral in (1.25). Here, the equation has been linearized in $\zeta(x)$ and its derivatives. Apart from that, one non-singular term of the order $\mathcal{O}(\sigma)$ has been neglected. It is hard to see, how a solution to equation (1.27) could be constructed. But *Fredholm's alternative* grants access to the eigenvalue spectrum without finding an explicit solution $Z(x)$: The equation has a solution, if and only if the kernel of the adjoint operator

1.2. Singular perturbation theory in the symmetric model

$\mathfrak{D}_2 + \tilde{\mathfrak{J}}_2$ is orthogonal to the inhomogeneity. It works for linear operators only. Let $\tilde{Z}^H(x)$ be a null eigenvector of the adjoint operator, i.e.

$$(\mathfrak{D}_2 + \tilde{\mathfrak{J}}_2) * \tilde{Z}^H(x) = 0 \,. \tag{1.29}$$

Then

$$\Lambda = \int_{-\infty}^{\infty} \frac{\tilde{Z}^H(x)}{(1+x^2)^{3/4}} = 0 \tag{1.30}$$

is a solvability condition. It turns out to be equivalent to the suppression of transcendentally small solution contributions at $x = 0$ in order to obtain a smooth dendrite tip ($\zeta'(0) = 0$). A WKB solution of (1.29) is inserted into (1.30) and the integration is carried out in the complex plane. A finite anisotropy of surface tension shifts the branch point of the integrand, around which the path of integration has to be taken. This leads to the existence of solutions. The root of Λ as a function of σ that corresponds to the stable solution, is found in the crossover range between oscillating and smooth behaviour of the integral. This yields the selected eigenvalue. All other roots are located at smaller σ, and they belong to thick, slow dendrites, the tips of which are unstable [82].

Alternatively, equation (1.25) instead of the Fredholm equation (1.30) can be solved. Brener and Mel'nikov constructed a solution [24] in a way similar to the *Kruskal-Segur method* [74, 75]. For large values of $|x|$, the derivatives of $\zeta(x)$ can be neglected. This leads to the outer equation

$$\sigma = \frac{(1+x^2)^{3/2}}{2\pi} \int_{-\infty}^{\infty} \frac{[\zeta(x') - \zeta(x)](x+x')}{\left[1 + \frac{1}{4}(x+x')^2\right](x-x')} \, \mathrm{d}x', \tag{1.31}$$

which can be solved exactly by means of the *Wiener-Hopf method*. The integral is calculated by the *method of residues*. The left hand side of (1.25) diverges at $x = \mp\mathrm{i}$ for $|\zeta'(x)| \ll 1$. Consequently, the

1. Basic concepts and approaches to dendritic growth

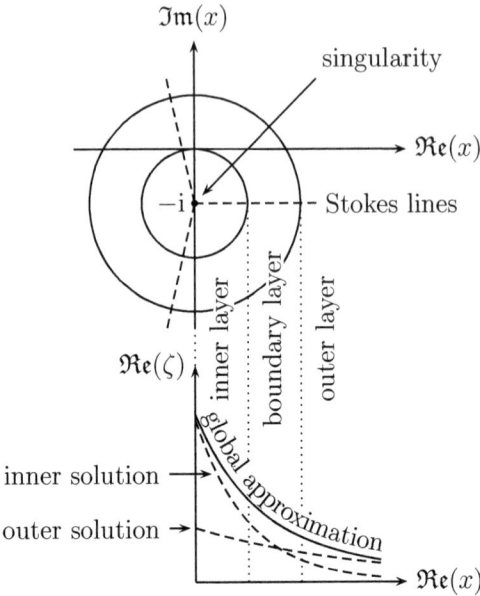

Figure 1.5.: Construction of a global approximation by asymptotic matching of the inner solution and the outer (WKB-)solution

next step is to derive a local form of (1.25) close to the singularity ($|x \pm \mathrm{i}| \ll 1$). Then, it must be possible that the solution of the inner equation asymptotically matches the solution of equation (1.31), as illustrated schematically in figure 1.5. This provides a solvability criterion. The anisotropy function reads

$$a(\theta) = 1 + \frac{2\beta}{(x - \zeta' - \mathrm{i})^2}, \quad (1.32)$$

where β is the strength of the crystalline surface tension anisotropy.

1.2. Singular perturbation theory in the symmetric model

The inner equation takes the form

$$\frac{d^2 F}{dz^2} - \frac{\sqrt{2}\lambda \tau_4^{7/2}}{\tau_4^2 - 2} = -1 \qquad (1.33)$$

with $x = i(1 - \sqrt{\beta}z)$, $\zeta(x) = \beta F(z)$ and $\tau_4 = z + F'(z)$. The stability parameter can be calculated from the nonlinear eigenvalue λ as $\sigma = \beta^{7/4}/\lambda$. The local equation (1.33) is solved numerically yielding the eigenvalue $\lambda = \beta^{7/4}/\sigma = 0.42$ [7]. Though, the discussion of solvability is fully analytical: The inner solution has its largest slopes on the three rays (*Stokes lines*) with $\arg z = 0, \pm\frac{4\pi}{7}$. Since one is interested in existing global solutions, the growth on these rays has to be suppressed. Otherwise a matching to the outer solution is not possible. Without anisotropy, there is no parameter λ, and one has only two integration constants, not enough to suppress growth on three rays. But with λ, there is a third parameter adjustable to produce a solution. This is how the stability parameter is selected mathematically in this framework. It shows, why there are no globally valid solutions at all in the case of isotropic surface tension. For $\beta = 0$, one has $\lambda = 0$, and the solution cannot be adjusted. Finally, the eigenvalue with the highest growth velocity corresponds to the only stable solution [24].

Both of the methods described above yield the scaling relations

$$V \propto \Delta^4 \beta^{\frac{7}{4}} \qquad (1.34a)$$
$$\rho \propto \Delta^{-2} \beta^{-\frac{7}{4}} \qquad (1.34b)$$

in the case of small P_c. In the limit of small undercooling, an explicit form of the selected growth velocity can be given:[3]

$$V = \frac{2D\Delta^4 \sigma}{\pi^2 d_0}. \qquad (1.35)$$

---

[3]Note, that if the one-sided model is used instead of the symmetric model, the velocity predicted in equation (1.35) is twice as large [88].

Due to the linearizations needed for Fredholm's alternative, only the second approach is considered to be rigorous. Anyway, both approaches share a crucial disadvantage: They can only be applied to linear field equations, because otherwise there is no Green's function. Especially the inclusion of convection into the theory is not possible with the methods described above.

1.3. Extended experimental and theoretical study of the selected pattern

1.3.1. Features of free dendritic growth

It is worth mentioning that the mathematical structure of diffusion-limited dendritic growth is the same as in viscous fingering, where a liquid of small viscosity is pushed into a liquid of high viscosity. In this respect, the pressure p is the relevant field quantity fulfilling a *Laplace equation*, and *Darcy's law* determines the flow velocity field. The *Saffman-Taylor instability* gives rise to the growth of finger-like patterns [62]. The problem was first solved in 1986 [110, 61, 34, 37]. Later on, the articles by Combescot et al. [33] and by Dorcey and Martin [38] presented the matter more elaborately. The analogy to dendritic growth was shown by Ben Amar and Brener [9]. The solution methods are the same, but the solution of the unperturbed problem has a very different structure. Apart from that, the sidebranches of Saffman-Taylor fingers can be unstable [82], whereas in free dendritic growth, sidebranches are stable and they are observed in every experiment. They also occur in directional solidification [47].[4] Sidebranches grow nearly perpendicularly to the growth direction of the

---

[4]In directional solidification, the shape of the growth cell is selected for a given wavelength of the pattern. But the wavelength can take values within a continuous band, and it is determined by the initial conditions of the problem.

1.3. Extended experimental and theoretical study of the selected pattern

main dendrite, and they migrate down its surface. They have no influence on the growth mode of the main dendrite [69], although LaCombe stated that there might be some kind of interaction between the sidebranches and the main dendrite's tip causing the growth velocity of the main dendrite to oscillate slightly around its mean corresponding to the stationary value [78]. These velocity oscillations are observed in some experiments. They can be tracked in the *Fourier transform* of the tip displacement measured as a function of time. Some modes are more pronounced than others. Several possible reasons for the oscillations have been proposed. In alloy solidification they are said to be a consequence of chemical gradients. Sawada et al. observed a dependence of the oscillation modes on the orientation of the crystal relative to the growth direction [105]. But in fact, the origin of the effect is still a mystery, and in the vast majority of the experiments, it is not observed.

Some of the hitherto cited experimental works were carried out in three dimensions. Ben Amar and Brener generalized the concept from section 1.2 to the three-dimensional case [8, 25]. Brener et al. also examined the problem of combined motion of melting and solidification fronts [27], triple junctions in three-phase systems [23, 20] and growth of non-reflection symmetric dendrites [26]. The latter case was investigated in more detail by Ben Amar and Brener, and the existence of double finger structures was predicted [9]. An example of these so called doublon structures is shown in figure 1.1d. They are observed for instance in the experiments of Stalder and Bilgram [116], and they can be formed, when an ordinary dendrite is destroyed by a tip-splitting instability. They can exist in systems with vanishing surface tension anisotropy. Doublons are the favourable growth mode at larger Δ. They are the building blocks of the *seaweed structure*, which grows faster than a compact dendritic crystal [21]. In three dimensions, the method described in subsection 1.2.2 also works. The crucial difference is that the 3D singular perturbation expansion breaks down far away from the tip. Thus, in the tail region the calculation has to be performed separately. The shape $x \sim y^{0.6}$ for the fins of the needle crystal is predicted [25] in contrast to the parabolic case in 2D. This result has been checked and verified by the group of Bilgram using xenon dendrites [116, 112, 113].

Dendritic growth also occurs in liquid crystals [93]. In this case, the diffusion coefficient D is an anisotropic tensor. For instance in a nematic liquid crystal, electrohydroconvection causes an instability selecting a continuous band of possible dendrite shapes [48]. The dynamics of a nematic-isotropic interface was investigated experimentally in [111]. In these systems, the substrate-nematic anchoring energy significantly enlarges the capillary length [54]. In the work of Börzsönyi et al. [17], the growth of a homeotropic (liquid crystal directors perpendicular to the 2D growth plane) smectic-B phase dendritic pattern into a planar (liquid crystal directors parallel to the 2D growth plane) nematic phase is analyzed. The twofold surface tension anisotropy strength ϵ_2 is determined from the comparison between experiments and phase-field simulations. It plays a crucial role, whether the nematic director is perpendicular or parallel to the growth direction. In case of the director being orientiated perpendicular to growth direction, the molecule incorporation into the smectic-B phase surface involves mainly twist, whereas in case of the director in growth direction the incorporation involves mainly splay and the energy of the kinetic changeover is remarkable. Thermal diffusion is less effective perpendicular to the nematic director, and growth happens faster in this direction [55]. This phenomenon is also referred to as "inverted growth".

Another interesting aspect to mention is rapid solidification [46], where the undercoolings and the growth velocities are large. It corresponds to the case of arbitrarily large growth Péclet numbers P_c, which was treated theoretically for instance in the work of Tanveer [117]. In the experiments of Funke et al. [45], undercoolings of up to 150 K are

1.3. Extended experimental and theoretical study of the selected pattern

realized. The experimenters manage to suppress nucleation, because they use ultra-high purity metals confined without walls in a magnetic field. The solidification front is tracked by a high-speed camera, because in samples sized in the millimeter range, the growth velocities are in the range of m/s instead of $\mu m/s$ which is the ordinary case. This is directly caused by the large undercooling. Since metals are optically non-transparent, the dendrite is visualized by a pyrometer. The released latent heat causes a measurable contrast at the two-phase boundary. The results agree quite well with microscopic solvability theory including kinetic effects. However, the agreement with theory might even be improved, when convection is included into the model, because there are flows caused by electromagnetic stirring in the metal droplet, which can be modeled as a forced flow.

Finally, it is worthwhile noting, that a finite thermal resistance (*Kapitza resistance*) at the two-phase boundary can influence dendritic growth. This was observed for instance by Rolley et al. [99], who carried out experiments using fluid ^{3}He at temperatures in the range $0.1 \ldots 0.32$ K. In this case, the two phases cannot be treated symmetrically anymore, and the model equations have to be altered (see section 4.2).

1.3.2. Previous analytical treatment of dendritic growth in convective systems

Achieving an analytical solution of the dendritic growth problem in a flow with as much rigour as possible is a main purpose of this work. Because of the high complexity, there have been only few attempts to treat solidification in convective systems theoretically without simulations. Schrage developed a simplified, three-dimensional model for natural convection, e. g. buoyancy-driven flows, from first principles based on Ivantsov's theory [107]. The convection is included via an additional energy transport term in the Stefan condition. This

1. Basic concepts and approaches to dendritic growth

term describes the heat transport away from the interface on a certain length scale by fluid flow. The model reaches quite formidable agreement of the *flow Péclet number*

$$P_f = \frac{\rho U}{D} \qquad (1.36)$$

as a function of the undercooling with existing experimental work [50, 101, 84, 118, 16]. However, the model does not provide a value of the stability parameter, since it is focused on the unstable solution of the problem without surface tension, comparable to other articles [6, 5]. There is no direct access to the flow velocity itself as a parameter. For parameter separation, Schrage takes the approximation $\sigma \approx 0.02$ from the orbital space flight microgravity experiment of Koss et al. [73]. Hence, the stability parameter is considered independent of any system parameters. σ is indeed independent of Δ at low undercoolings, but the effect of convection on σ is of particular interest here.

Saville and Beaghton showed that a parabolic similarity solution exists in the case of zero surface tension if a forced *Oseen flow* is taken into account [104]. They predict the dependence of the growth Péclet number P_c on the flow for fixed undercoolings. Two-dimensional calculations on growth mode selection using the Oseen flow approximation were presented by Bouissou and Pelcé [19]. Their publication received much attention. It is based on a linear stability analysis and predicts

$$\frac{1}{\sigma} = \frac{8}{\beta^{7/4}} \left[1 + \tilde{b} \left(\frac{\tilde{a} U d_0}{\beta^{3/4} \rho V} \right)^{\frac{11}{14}} \right] \qquad (1.37)$$

for the stability parameter as a function of the forced flow velocity U defined in section 1.1. $\tilde{a}$ is defined later in (3.39), and $\tilde{b}$ is a numerical constant. The authors compare the result (1.37) with their own experiments [18], where a linear increase of $1/\sigma$ as a function of U is measured. It is then concluded, that the agreement is fairly good, because a scale exponent of $\frac{11}{14}$ is hard to distinguish from 1

1.3. Extended experimental and theoretical study of the selected pattern

regarding experimental accuracy. But the deviations may also be a consequence of the linearizations made in the calculations. In particular, the operator in the Fredholm-type solvability condition uses only an approximate form of the curvature. The linearizations are a bad approximation close to the singular point of the problem. Apart from that, the employed form of the Oseen flow is not a good approximation in 2D [10]. More precise results would be worth some effort. A comparison to the results of Schaefer and Coriell could be made, who found that very small ethanol concentrations in succinonitrile are sufficient for an onset of thermosolutal convection and a change of the shape and the stability properties of the interface [106]. Apart from that, especially the newer convective experiments of Lee [84] or Emsellem [42] should be compared to theories.

Sekerka et al. described the convective effects on dendritic growth using a stagnant film model [108], which is also focused on a buoyancy-driven flow just as the model of Schrage. Furthermore, there has been a work by Li and Beckermann on analytical modeling of dendritic growth in thermosolutal melt convection driven by chemical diffusion [85]. Other theoretical publications on dendritic growth in convective systems are about simulations. For some newer results see for example [120, 119, 80, 86]. In this respect, the phase-field model of Karma and Rappel has become a maintained standard [67]. It had first been applied to the flowless case of dendritic growth in two and three dimensions [68]. The phase of the material is represented by a field variable with values ranging from -1 in the liquid phase to $+1$ in the solid phase. It enters the defining equations and varies fast but not infinitely fast in the two-phase boundary layer. Such phase field simulations are also a suitable tool for examining directional solidification [1, 60, 70, 36].

1. Basic concepts and approaches to dendritic growth

1.3.3. Uncertainty about crystalline anisotropy strength measurements in relevant materials

Any theory on dendritic growth needs material parameters as input to become comparable to experiments. A "relevant" material should behave metal-like regarding its solidification properties, because predictions about metals are desirable due to their wide application range. However, tracking the growing dendrite with a camera in an experiment is a convenient method, and this is only possible, if the used substance is optically transparent. *Plastic crystals* generally fulfill these two criteria. Two substances of this type are for instance *succinonitrile* (body-centered cubic lattice) and *pivalic acid* (face-centered cubic lattice).

The strength of the fourfold crystalline anisotropy of capillary effects β is usually provided by experimental measurements. As mentioned in subsection 1.2.2, the stability parameter scales as $\sigma \sim \beta^{7/4}$. This rather strong dependence will lead to notable deviations in the prediction of σ, if β is not measured accurately enough. In general one has

$$\tilde{\gamma}(\theta) = \frac{\gamma(\theta)}{\gamma_0} = 1 + \sum_{k=1}^{\infty} \epsilon_{2k} \cos(2k\theta) \tag{1.38}$$

for the anisotropic surface tension, where γ_0 denotes the average over all angles θ. In substances with cubic lattice structures the anisotropy is essentially fourfold, i.e. the ϵ_4-term constitutes the dominant contribution. Physically, this means that there are several coldest points on the solid surface. This has a stabilizing effect as already explained in section 1.1. From (1.38), one can calculate the anisotropy function $a(\theta)$:

$$a(\theta) = \tilde{\gamma}(\theta) + \tilde{\gamma}''(\theta) = 1 - \underbrace{15\,\epsilon_4}_{=\beta} \cos(4\theta) \,. \tag{1.39}$$

$\gamma(\theta) + \gamma''(\theta)$ is also referred to as surface stiffness. Hence, an error in ϵ_4, which is usually the measured quantity, is amplified by a factor

1.3. Extended experimental and theoretical study of the selected pattern

Table 1.1.: Reported values of the fourfold crystalline anisotropy strength of capillary effects ϵ_4 and $\beta = 15\epsilon_4$ from different experiments carried out with pivalic acid

ϵ_4	β	citation
0.006	0.09	[39]
0.025	0.375	[91]
0.050	0.75	[52]

of 15 when calculating β, even though a relative error is unchanged. Experimental measurements of the surface energy γ and its anisotropy strength ϵ_4 for polar liquids and many other substances can be found in [92, 32]. The most common measurement method (but not the only one existing) is to determine the angular dependence of the distance between the center and the surface of a crystal in its equilibrium shape. The parameters are then calculated by adapting a Wulff construction [65, 109]: The anisotropy coefficients ϵ_{2k} are the amplitudes of the angular dependence function's Fourier modes. Yet, different values of ϵ_4 for pivalic acid have been measured (see table 1.1). In the work of Dougherty [39], the most details about the experimental procedure were given, and the growth samples were equilibrated for the longest time of all three reports listed in the table. This value yields the best agreement between the theory presented here and experiments. In the work of Muschol et al. [89], a remarkable deviation between the measured value of σ and the value predicted by microscopic solvability theory is stated, especially for pivalic acid. But in the publication, the value predicted by microscopic solvability theory is calculated using their measured value of ϵ_4 according to the above mentioned scaling relation. The authors themselves state that the measured value of ϵ_4 is quite large and unreliable. In fact, the crystals were not measured at constant size and hence, the observed growth shapes were not necessarily identical to equilibrium shapes. The relatively low experimental values of the stability parameter together with the relatively high anisotropies from the last two lines in table 1.1 suggest that the ordinary microscopic solvability theory for calculating σ from ϵ_4 might not work for pivalic acid. In fact, pivalic acid crystals grow strongly anisotropic and with a tendency to facets. This indicates however, that microscopic solvability theory could still yield accurate results, if kinetic effects, which are assumed to play an important role in pivalic acid, were taken into account. This is investigated in section 4.5 of this work.

2. Advanced approach for dendritic growth in nonlinear systems

In this chapter, the solution concepts and methods used in this work are introduced. They are shown in a basal manner, and simple but yet very general examples are used for illustration. The schemes presented in this chapter are used extensively in the succeeding chapters, and they allow to conveniently shorten the exhibited calculus at most points.

2.1. Conformal parabolic coordinates and non-dimensionalization

From here on, most calculations will be carried out in a coordinate system adapted to the problem. We use conformal parabolic coordinates in the xy-plane:

$$x = \eta \xi, \tag{2.1a}$$

$$y = \frac{1}{2}\left(\eta^2 - \xi^2\right). \tag{2.1b}$$

With $\xi \in [-\infty, \infty]$ and $\eta \in [0, \infty]$, the whole xy-plane is covered. A line of constant η is a concave parabola, a point on which is specified by ξ. The boundary conditions are easy to handle in these coordinates, and describing the Ivantsov isotherms is almost trivial. Ananth and Gill first used these coordinates [10] to construct a similarity solution for the temperature field in convective dendritic growth. The

2. Advanced approach for dendritic growth in nonlinear systems

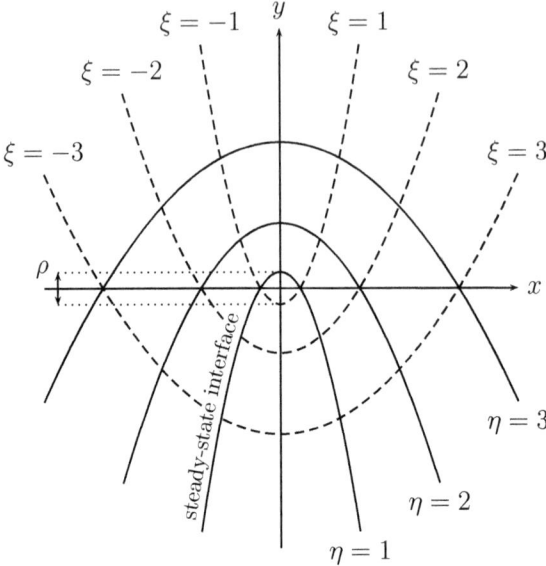

Figure 2.1.: Conformal parabolic coordinates, $\eta \in [0, \infty]$ specifies a concave parabola, $\xi \in [-\infty, \infty]$ specifies a point on the parabola

coordinates (2.1) are chosen, so that they have a locally orthogonal basis. This is not necessarily mandatory, but it leaves us with much handier expressions. The coordinate set is called "conformal", because it fulfills $|\partial \vec{r}/\partial \xi| = |\partial \vec{r}/\partial \eta|$. The normalized basis vectors are:

$$\vec{e}_x = \frac{1}{\sqrt{\xi^2 + \eta^2}} (\eta \vec{e}_\xi + \xi \vec{e}_\eta) \quad \vec{e}_y = \frac{1}{\sqrt{\xi^2 + \eta^2}} (\eta \vec{e}_\eta - \xi \vec{e}_\xi) \quad (2.2a)$$

$$\vec{e}_\xi = \frac{1}{\sqrt{\xi^2 + \eta^2}} (\eta \vec{e}_x - \xi \vec{e}_y) \quad \vec{e}_\eta = \frac{1}{\sqrt{\xi^2 + \eta^2}} (\xi \vec{e}_x + \eta \vec{e}_y) \; . \quad (2.2b)$$

We find the nabla operator and the Laplace operator:

$$\vec{\nabla} = \frac{1}{\sqrt{\xi^2 + \eta^2}} \left[\vec{e}_\xi \frac{\partial}{\partial \xi} + \vec{e}_\eta \frac{\partial}{\partial \eta} \right] , \quad (2.3a)$$

2.1. Conformal parabolic coordinates and non-dimensionalization

$$\Delta = \frac{1}{\xi^2 + \eta^2} \left[\frac{\partial^2}{\partial \xi^2} + \frac{\partial^2}{\partial \eta^2} \right]. \tag{2.3b}$$

Let $\eta_s(\xi) - \eta = 0$ describe the interface. The position vector at the interface is

$$\vec{r} = \eta_s \xi \vec{e}_x + \frac{1}{2}\left(\eta_s^2 - \xi^2\right) \vec{e}_y \tag{2.4}$$

and the differential line element is

$$\mathrm{d}s = \sqrt{\mathrm{d}x^2 + \mathrm{d}y^2} = \sqrt{(\eta_s^2 + \xi^2)(1 + \eta_s'^2)}\, \mathrm{d}\xi. \tag{2.5}$$

From the condition

$$0 = \frac{\mathrm{d}\vec{r}}{\mathrm{d}s} \cdot \vec{n} = \vec{t} \cdot \vec{n}$$

we derive the unit tangent vector and the unit normal vector to the interface

$$\vec{t} = \frac{(\eta_s + \xi \eta_s')\vec{e}_x + (\eta_s \eta_s' - \xi)\vec{e}_y}{\sqrt{(\eta_s^2 + \xi^2)(1 + \eta_s'^2)}} = \frac{\vec{e}_\xi + \eta_s' \vec{e}_\eta}{\sqrt{1 + \eta_s'^2}} \tag{2.6a}$$

$$\vec{n} = \frac{(\eta_s + \xi \eta_s')\vec{e}_y - (\eta_s \eta_s' - \xi)\vec{e}_x}{\sqrt{(\eta_s^2 + \xi^2)(1 + \eta_s'^2)}} = \frac{\vec{e}_\eta - \eta_s' \vec{e}_\xi}{\sqrt{1 + \eta_s'^2}} \tag{2.6b}$$

where $\vec{n}$ is chosen to point into the liquid. The curvature is

$$\kappa = -\vec{n} \cdot \frac{\mathrm{d}^2 \vec{r}}{\mathrm{d}s^2} = -\frac{1}{\sqrt{\xi^2 + \eta_s^2}} \left[\frac{\eta_s''}{(1 + \eta_s'^2)^{\frac{3}{2}}} + \frac{\eta_s' \xi - \eta_s}{(\xi^2 + \eta_s^2)\sqrt{1 + \eta_s'^2}} \right]. \tag{2.7}$$

κ, as noted here, is positive and takes its maximum value at $\xi = 0$. We will consider a twofold or a fourfold crystalline anisotropy:

$$a_2(\theta) = 1 - \frac{\beta_2}{2}\cos(2\theta) = 1 + \frac{\beta_2}{2}\left(1 - 2\cos^2\theta\right) \tag{2.8a}$$

$$a_4(\theta) = 1 - \beta_4 \cos(4\theta) = 1 - \beta_4\left(1 - 8\cos^2\theta \sin^2\theta\right) \tag{2.8b}$$

where

$$\cos\theta = \vec{n} \cdot \vec{e}_y = \frac{[\xi \eta_s]'}{\sqrt{(\xi^2 + \eta_s^2)(1 + \eta_s'^2)}}, \tag{2.9a}$$

2. Advanced approach for dendritic growth in nonlinear systems

$$\sin\theta = \vec{n}\cdot\vec{e}_x = \frac{\xi - \eta_s\eta_s'}{\sqrt{(\xi^2+\eta_s^2)(1+\eta_s'^2)}}. \tag{2.9b}$$

Using this, we find the anisotropy functions expressed in parabolic coordinates:

$$a_2(\theta) = 1 - \frac{\beta_2}{2}\frac{[\xi\eta_s]'^2 - (\eta_s\eta_s' - \xi)^2}{[\xi\eta_s]'^2 + (\eta_s\eta_s' - \xi)^2}, \tag{2.10a}$$

$$a_4(\theta) = 1 - \beta_4\left[1 - 8\frac{(\xi - \eta_s\eta_s')^2(\eta_s + \xi\eta_s')^2}{(\xi^2+\eta_s^2)^2(1+\eta_s'^2)^2}\right]. \tag{2.10b}$$

We define the stream function ψ by

$$\vec{w} = \vec{\nabla}\times(\psi\vec{e}_z) \tag{2.11}$$

yielding

$$\vec{w} = \frac{1}{\xi^2+\eta^2}\begin{vmatrix} \sqrt{\xi^2+\eta^2}\,\vec{e}_\xi & \sqrt{\xi^2+\eta^2}\,\vec{e}_\eta & \vec{e}_z \\ \frac{\partial}{\partial\xi} & \frac{\partial}{\partial\eta} & \frac{\partial}{\partial z} \\ 0 & 0 & \psi \end{vmatrix} = \frac{1}{\sqrt{\xi^2+\eta^2}}\left(\vec{e}_\xi\psi_\eta - \vec{e}_\eta\psi_\xi\right)$$

in parabolic coordinates. By this choice, condition (1.11) (div $\vec{w} = 0$) is automatically fulfilled. I.e. $\vec{w}$ is going to describe an incompressible flow for any function $\psi(\xi,\eta)$. This simplification works in 2D, but not necessarily in 3D.

As noted at the end of section 1.1 in the preceding chapter, we are generally interested in steady-state solutions with stationary growth velocity V. Thus, we switch to a moving frame of reference,

$$\vec{r} \to \vec{r} + Vt\vec{e}_y$$
$$\vec{w} \to \vec{w} + V\vec{e}_y$$

resulting in

$$\frac{\partial}{\partial t} \to -V(\vec{e}_y\cdot\vec{\nabla}) = V\left(\xi\frac{\partial}{\partial\xi} - \eta\frac{\partial}{\partial\eta}\right).$$

In addition to that, we apply some scalings, rendering the model equations dimensionless:

- We measure all lengths in units of the dendrite tip curvature radius ρ:
$$x, y \to \rho x, \rho y, \ \kappa \to \frac{\kappa}{\rho}.$$

- The temperature is made dimensionless and measured in units of the growth Péclet number $P_c = \frac{\rho V}{D}$:
$$T \to T_M + P_c \frac{L}{c} T.$$

- The stream function is measured in units of the diffusion coefficient:
$$\psi \to D\psi, \ \vec{w} \to \frac{D}{\rho}\vec{w}.$$

- The time is rescaled:
$$t \to \frac{\rho^2}{D} t.$$

The resulting dimensionless model equations will be given at the beginning of each section, adapted to the respective type of problem.

2.2. Demonstration of the analytical part of the method by application to the symmetric model

The construction of a system analogue to (1.31)-(1.33) selecting the growth mode in the case of nonlinear field equations is not possible using Green's functions. The careful application of an old approximative scheme à la *Zauderer* [125] can lead to success: The asymptotic decomposition of partial differential equations on a slowly varying scale can

2. Advanced approach for dendritic growth in nonlinear systems

also be applied to nonlinear problems. In combination with asymptotic matching in the complex plane, it works where classical microscopic solvability theory cannot work. This particular combination of tools was introduced by Thomas Fischaleck (now named T. Grillenbeck) [44].

In this section, the scheme is demonstrated by application to the flowless dendritic growth problem, which was already treated in section 1.2. This will not provide any new results, but the reader may get an idea of the method and its power, because the results presented in section 1.2 can be derived with much less analytical effort.

Using the scalings from the preceding section, the temperature diffusion equation in the moving frame of reference is

$$T_{\xi\xi} + T_{\eta\eta} + P_c(\eta T_\eta - \xi T_\xi) = 0 \tag{2.12}$$

in the liquid and in the solid domains in conformal parabolic coordinates. The Ivantsov parabola now reads $\eta_s = 1$. The temperature field and the interface position are expanded about the Ivantsov solution:

$$T = T^{\text{Iv}} + T^1 + \ldots \tag{2.13a}$$
$$\eta_s = 1 + h(\xi) + \ldots \tag{2.13b}$$

The interface is shifted by the small term $h(\xi)$ and is no longer perfectly parabolic. The shape function $h(\xi)$ is to be determined. Since (2.12) is linear, it also applies for T^1 and the superscript 1 is dropped right away. Using the definitions

$$\vec{\vartheta} = \begin{pmatrix} T_\xi \\ T_\eta \end{pmatrix} \tag{2.14}$$

$$A = \begin{pmatrix} 0 & 1 \\ -1 & 0 \end{pmatrix} \qquad B_0 = \begin{pmatrix} -P_c\xi & P_c\eta \\ 0 & 0 \end{pmatrix} \tag{2.15}$$

2.2. Demonstration of the analytical part of the method by application to the symmetric model

the field equation is written as a first order system. This step is useful before continuing with the asymptotic decomposition procedure:

$$\vec{\vartheta}_\xi + A\vec{\vartheta}_\eta + B_0\vec{\vartheta} = 0. \tag{2.16}$$

The variable $\vec{\vartheta}$ shall be written as a linear combination of eigenvectors of A. A has the eigenvalues i and $-$i with eigenvectors

$$\vec{r}_1 = \begin{pmatrix} -i \\ 1 \end{pmatrix} \quad \text{and} \quad \vec{r}_2 = \begin{pmatrix} i \\ 1 \end{pmatrix} \tag{2.17}$$

respectively. Now we write the asymptotic decomposition

$$\vec{\vartheta} = M\vec{r}_1 + N\vec{r}_2 \tag{2.18}$$

of the temperature field. The coefficients M and N have to be determined as spatial functions. In the original Zauderer method, one of the two terms in the linear combination (2.18) is multiplied by a factor ε in order to emphasize that the contribution is small compared to the other one [125]. A priori we do not know, which term that is. It shall be investigated here. (2.18) is inserted into (2.16):

$$(M_\xi + iM_\eta)\vec{r}_1 + (N_\xi - iN_\eta)\vec{r}_2 + B_0 M\vec{r}_1 + B_0 N\vec{r}_2 = 0. \tag{2.19}$$

Again, this equation applies in both phases. The next step is to project (2.19) onto the invariant subspaces of A. We search for projection operators defined by $P_i \vec{r}_j = \delta_{ij}\vec{r}_i$ (no summation convention):

$$P_1 = \frac{1}{2}\begin{pmatrix} 1 & -i \\ i & 1 \end{pmatrix}, \qquad P_2 = \frac{1}{2}\begin{pmatrix} 1 & i \\ -i & 1 \end{pmatrix}. \tag{2.20}$$

Applying P_1, P_2 to (2.19) we find

$$M_\xi + iM_\eta = \frac{P_c}{2}(\xi - i\eta)M - \frac{P_c}{2}(\xi + i\eta)N, \tag{2.21a}$$

$$N_\xi - iN_\eta = -\frac{P_c}{2}(\xi - i\eta)M + \frac{P_c}{2}(\xi + i\eta)N. \tag{2.21b}$$

2. Advanced approach for dendritic growth in nonlinear systems

Here, the formulas

$$P_1 B_0 \vec{r}_1 = -\frac{P_c}{2}(\xi - i\eta)\,\vec{r}_1 \qquad P_1 B_0 \vec{r}_2 = \frac{P_c}{2}(\xi + i\eta)\,\vec{r}_1 \qquad (2.22a)$$

$$P_2 B_0 \vec{r}_1 = \frac{P_c}{2}(\xi - i\eta)\,\vec{r}_2 \qquad P_2 B_0 \vec{r}_2 = -\frac{P_c}{2}(\xi + i\eta)\,\vec{r}_2 \qquad (2.22b)$$

were used. (2.21a) and (2.21b) simplify in the limit $P_c \to 0$:

$$M_\xi + i M_\eta = 0\,, \qquad (2.23a)$$
$$N_\xi - i N_\eta = 0\,. \qquad (2.23b)$$

To find out, which one of the functions M and N is small compared to the other one, the Gibbs-Thomson condition is differentiated along the interface:

$$\frac{\mathrm{d}T^s}{\mathrm{d}\xi} = T_\xi^s + h' T_\eta^s = -\frac{1}{2}\sigma[\kappa a(\theta)]'\,. \qquad (2.24)$$

Regaring the definition (2.14) of $\vec{\vartheta}$ and its decomposition (2.18), the derivatives of T^s are

$$T_\xi^s = i\,(N^s - M^s)\,, \qquad T_\eta^s = N^s + M^s\,. \qquad (2.25)$$

and this is inserted into (2.24):

$$(1 + ih') M^s - (1 - ih') N^s = -\frac{i}{2}\sigma[\kappa a(\theta)]'\,. \qquad (2.26)$$

The curvature from (2.7) on the right hand side of (2.26) diverges at $\xi = \pm i$ for $h \ll 1$. We wish to analyze (2.26) in the vicinity of the singularity at $\xi = -i$, and we follow Fischaleck [43]. The general solutions of (2.23a) and (2.23b) are of the form

$$M = M\bigl(\xi + i(\eta - 1)\bigr)\,, \qquad N = N\bigl(\xi - i(\eta - 1)\bigr)\,. \qquad (2.27)$$

Consider the derivatives (2.25) with the special argument forms from (2.27) at $(\xi, \eta) = (0, 0)$:

$$T_\xi^s = i\bigl(N^s(i) - M^s(-i)\bigr)\,, \qquad T_\eta^s = N^s(i) + M^s(-i)\,. \qquad (2.28)$$

2.2. Demonstration of the analytical part of the method by application to the symmetric model

These derivatives cannot diverge here, because $(\xi, \eta) = (0,0)$ is just an ordinary point in the solid. Consequently, $M^s(-i)$ and $N^s(i)$ are finite. Thus, when writing (2.26) at $\xi = -i$, the singularity can only be compensated by $N^s(-i)$. We conclude $N^s \gg M^s$ in the vicinity of $\xi = -i$, and an analogue argumentation leads to $N^l \ll M^l$ at this location. The perturbative scheme arising from Zauderer decomposition corresponds to an expansion about the analytic continuation of the Ivantsov solution in the vicinity of the singularity (at $\xi = -i$ in this work).

Note, that the same statement can be made about the temperature itself instead of its derivatives: For $P_c \to 0$, the diffusion equation (2.12) reads

$$T_{\xi\xi} + T_{\eta\eta} = 0. \tag{2.29}$$

It is of Laplace type. Consider the two equations

$$(\partial_\xi + i\partial_\eta) T_a = 0, \tag{2.30a}$$
$$(\partial_\xi - i\partial_\eta) T_b = 0. \tag{2.30b}$$

Obviously, solutions of (2.30a) have the form $T_a = T_a(\xi + i(\eta - 1))$, and solutions of (2.30b) have the form $T_b = (\xi - i(\eta - 1))$. Because of $\partial_{\xi\xi} + \partial_{\eta\eta} = (\partial_\xi + i\partial_\eta)(\partial_\xi - i\partial_\eta)$, any solutions of (2.30a) and (2.30b) are also solutions of equation (2.29), and a superposition of T_a and T_b is a general solution for the temperature field:

$$T^l(\xi,\eta) = T_a^l(\xi + i(\eta - 1)) + T_b^l(\xi - i(\eta - 1)), \tag{2.31a}$$
$$T^s(\xi,\eta) = T_a^s(\xi + i(\eta - 1)) + T_b^s(\xi - i(\eta - 1)). \tag{2.31b}$$

As explained above, T_b^l and T_a^s may be neglected close to $\xi = -i$. Then T^l is a solution of (2.30a) and T^s is a solution of (2.30b):

$$iT_\xi^l = T_\eta^l, \tag{2.32a}$$
$$-iT_\xi^s = T_\eta^s. \tag{2.32b}$$

2. Advanced approach for dendritic growth in nonlinear systems

After these considerations, the decomposition (2.18) can be rewritten in an appropriate manner:

$$\vec{\vartheta} = M\vec{r}_1 + \varepsilon N \vec{r}_2, \tag{2.33a}$$

$$\vec{\vartheta}^s = N^s \vec{r}_2. \tag{2.33b}$$

(2.33a) applies in the liquid. The superscript l is omitted from now on. Since M^s is expected to be much smaller than N^s in the vicinity of $\xi = -\mathrm{i}$, the M^s-term was neglected completely in (2.33b). We could have proceeded the same way with N in (2.33a). That would lead to the same result for the equation determining the shape correction $h(\xi)$. Instead, the N-term is kept because in the next chapter, the field equation in the liquid will not be a Laplace equation anymore due to convection terms. Then, N will be crucial for the calculation of first order flow contributions, and it shall be demonstrated here, how to apply the method with three decomposition coefficients instead of only two. However, N is multiplied by a factor ε to indicate that it is small compared to M. In case of equation (2.29), the first order Zauderer scheme leads to an exact solution. However, if the field equation is not a Laplace equation, the neglect of M^s or N is not possible in general, and the presented scheme will only provide an approximate solution.

We assume the solution to be slowly varying with respect to (ξ, η). Therefore, the scale transformation

$$\xi, \eta \to \varepsilon \xi, \varepsilon \eta \tag{2.34}$$

is made. It emphasizes the principal part of the field equations.[1] Inserting (2.33a)-(2.33b) into the field equation (2.16) at $P_c = 0$ ($\Rightarrow B_0 = 0$), we get

$$(M_\xi + \mathrm{i} M_\eta)\, \vec{r}_1 + \varepsilon\, (N_\xi - \mathrm{i} N_\eta)\, \vec{r}_2 = 0 \quad \text{in the liquid} \tag{2.35a}$$

---

[1] The small parameter ε does not need to be chosen arbitrarily. Within the theory it is $\sigma^{2/7} \ll 1$.

2.2. Demonstration of the analytical part of the method by application to the symmetric model

$$\left(N^s_\xi - \mathrm{i} N^s_\eta\right) \vec{r}_2 = 0 \quad \text{in the solid.} \quad (2.35b)$$

One has to be careful about which terms to neglect, because after returning to the original scale, a factor ε will be reabsorbed into any coefficients linear in ξ or η. Contributions from a desired effect such as convection must be kept at least in their lowest occuring order. The intention is to determine the coefficients M, N and N^s as the dependent functions. After the projection part of the method and returning to the original scale, we find (2.23a)-(2.23b) in the liquid, supplemented by

$$N^s_\xi - \mathrm{i} N^s_\eta = 0 \quad (2.36)$$

in the solid. Note, that equations (2.23a)-(2.23b) have the same mathematical structure as (2.30a)-(2.30b). These equations can be solved for M, N and N^s subsequently. The boundary conditions of the problem have to be written in their dimensionless form in parabolic coordinates and they have to be decomposed, too. This has already been carried out in (2.26) exemplarily for the Gibbs-Thomson condition. An explicit exhibition of those calculations is postponed to the next chapter. Here the solution

$$N = \frac{\mathrm{i}}{2}\left[\frac{\sigma[\kappa(\xi)a(\theta)]'}{1-\mathrm{i}h'(\xi)} - \frac{[(1-\mathrm{i}\xi)h(\xi)]'}{1-\mathrm{i}h'(\xi)}\right] \quad (2.37)$$

is just written down where the interface conditions have already been applied. In (2.37), ξ has to be considered as a complex variable. For the far field boundary condition (1.10a) to hold, we must have $N \to 0$ as $\eta \to \infty$, hence

$$\sigma \kappa a(\theta) = (1 - \mathrm{i}\xi) h(\xi). \quad (2.38)$$

This equation can also be derived from a parabolic-coordinate version of equation (1.31) using the residues method. Thus, the results from [24] have been reproduced by means of asymptotic decomposition in just a few simple steps without using Green's functions, Bessel functions or residuals. One could now continue with the asymptotic

2. Advanced approach for dendritic growth in nonlinear systems

matching procedure. (2.38) is valid close to the singular point located at $\xi = -i$. If we had wished to expand in the vicinity of the other singularity at $\xi = i$, we would have had to use a different ansatz instead of (2.33a)-(2.33b). The results would have been almost the same. Only the minus sign in the brackets on the right hand side of (2.38) would be replaced by a plus sign. At $\xi = i$, one finds nothing but the complex conjugate problem, which does not contain additional information.

Zauderer's decomposition scheme seems to have passed largely into oblivion, although it is a powerful tool. For instance, it should also be a suitable approach to the complex *Ginzburg-Landau model* of superconductivity, which can be reduced to a free boundary problem [31].

2.3. Numerical treatment of the local equation

The local equations in the sense of (1.33) appearing in this work have the form

$$\tilde{\kappa}a(\theta) = \phi t + P \int^{t} \int^{t'}_{\infty} \tilde{l}\left(\phi(t'), \dot{\phi}(t'), \phi(t''), \dot{\phi}(t''), t', t''\right) dt'' dt', \quad (2.39)$$

and they are valid on an asymptotically small disk around the chosen singularity in the complex plane. In some cases, the right hand side is altered. The equations have to be treated numerically to find the eigenvalue σ. Here, t is the independent complex variable, and $\phi(t)$ is the dependent function solving the integro-differential equation (2.39). ϕ is the local form of the shape correction function h after stretching transformation close to the singularity. $\tilde{\kappa}$ and $a(\theta)$ have to be written in terms of ϕ and t. P is a dimensionless number to be specified later. It can be considered as a given parameter.

2.3. Numerical treatment of the local equation

We wish to obtain the function $\sigma(P)$ numerically from equation (2.39). This requires some effort, since we have to solve a second order integro-differential equation in the complex plane. Most of the common algorithms for ordinary differential equations require a real first order system. For this purpose, we write

$$x_1(t) = \phi(t), \tag{2.40a}$$
$$x_2(t) = \dot{\phi}(t) = \dot{x}_1(t). \tag{2.40b}$$

Furthermore, we write the integral in (2.39) as an additional dependent function:

$$x_3(t) = \int_\infty^t \int^{t'} \tilde{l}\left(\phi(t'), \dot{\phi}(t'), \phi(t''), \dot{\phi}(t''), t', t''\right) \mathrm{d}t'' \mathrm{d}t'. \tag{2.41}$$

In this manner, we avoid differentiating, which would leave us with a more complicated third order equation. The next step is to rearrange (2.39) until $\dot{x}_2(t) = \ddot{\phi}(t)$ appears solitary on the left hand side. Then, the right hand side is named $f(\{x_i\}, t)$. Introducing the abbreviations

$$\begin{aligned} p_1 &= x_1 t + P x_3 & p_2 &= t + x_1 \\ p_3 &= 1 - x_2 & p_4 &= 1 + x_2 \end{aligned} \tag{2.42}$$

turns out to reasonably simplify things a bit in general. Now the first order system can be written:

$$\dot{x}_1 = x_2 \tag{2.43a}$$
$$\dot{x}_2 = f(\{x_i\}, t) \tag{2.43b}$$
$$\dot{x}_3 = g(\{x_i\}, t). \tag{2.43c}$$

The functions $f(\{x_i\}, t)$ and $g(\{x_i\}, t)$ are given by

$$f(\{x_i\}, t) = \ddot{\phi} \tag{2.44a}$$

2. Advanced approach for dendritic growth in nonlinear systems

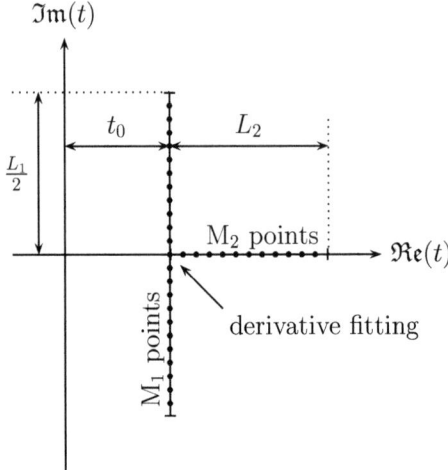

Figure 2.2.: Numerical cross-linked integration scheme for local equations in the complex plane adapted from Tanveer [117]

$$g\left(\{x_i\}, t\right) = -\int_t^\infty \tilde{l}\left(\phi(t), \dot{\phi}(t), \phi(t'), \dot{\phi}(t'), t, t'\right) dt' \qquad (2.44\text{b})$$

respectively. If we differentiated, the function $f\left(\{x_i\}, t\right)$ would contain many more terms.

Keep in mind that we deal with an eigenvalue problem. For given P, there are solutions only for isolated values of σ. The only stable solution corresponds to the largest value of σ or to the highest growth velocity V. The strategy to find σ is adapted from Tanveer [117]. Figure 2.2 visualizes the integration scheme. First, the system of equations (2.43) is solved on a line with length L_1 parallel to the imaginary axis positioned symmetrically with respect to the real axis, i.e. the independent variable has a constant non-zero real part $t_0 \in \mathbb{R}$. The boundary conditions at $t = t_0 \pm \mathrm{i}\frac{L_1}{2}$ follow from asymptotic anal-

2.3. Numerical treatment of the local equation

ysis of (2.39) and from symmetry requirements. It is explained in more detail at the end of this section. Once a solution on this line parallel to the imaginary axis is found, the values of x_1 and x_3 at t_0 are stored. Subsequently, the problem is solved on the real axis on a line with length L_2 starting at t_0. Here, the stored values of x_1 and x_3 are used as lower boundary conditions, since we look for analytic solutions, which have to be continuous in all derivatives. The upper boundary condition at $t = t_0 + L_2$ is obtained by asymptotic analysis again. Then the difference between x_2 at the crossing point of both lines is calculated and driven to zero by an external root-finder subroutine, which may vary σ as independent variable. The implementation ensures that the solution fulfills the selection criterion once σ becomes equal to the eigenvalue. For the results presented in this work, $L_1 = 20$, $L_2 = 10$ and $t_0 = 2.0$ were used, which turned out to be sufficiently large.

Several algorithms have been tried to solve the equation. A *shooting method* did not succeed, neither using 6th order *Runge-Kutta integration* nor using a *Kaps-Rentrop*-like subroutine for stiff equations. The commercial ODE integrators of **MATLAB** also failed. Finally, a solution was found using the powerful relaxation method. So to speak, we followed *William H. Press'* famous directive *"First shoot, then relax!"* from the book *"Numerical Recipies"* [96], which has become a standard in numerical computation. The relaxation method determines the solution by starting with an initial guess and improving it, iteratively. As the iterations improve the solution, the result is said to relax to the true solution [96]. In preparation of the relaxation procedure, the integration interval is discretized, and at each mesh point, (2.43) turns into a set of finite difference equations. We must supply an initial guess for the solution, which is taken from asymptotic analysis of equation (2.39). The guess is "relaxed" towards the solution step by step at each discretization point until a sophisticated error criterion is fulfilled.

2. Advanced approach for dendritic growth in nonlinear systems

The integral $g(\{x_k\},t)$ constitutes a notable problem. Methods dealing with integro-differential equations are rare and usually quite advanced [96]. We need to evaluate the integral at each mesh point during each relaxation iteration. How is that possible? Fortunately, the integration boundaries of g are compatible with the relaxation interval, and the relaxation method provides a matrix containing the current versions of x_1 and x_2 at each iteration step. From this, g is calculated by the trapezoidal rule.

In the following, some implementation details parallel to the imaginary axis are given as an example for the reader to get an idea of the steps to be taken. We write

$$x_{1,k} = \frac{1}{2}(y_{1,k} + y_{1,k-1}) + \frac{i}{2}(y_{2,k} + y_{2,k-1}) \qquad (2.45\text{a})$$

$$x_{2,k} = \frac{1}{2}(y_{3,k} + y_{3,k-1}) + \frac{i}{2}(y_{4,k} + y_{4,k-1}) \qquad (2.45\text{b})$$

$$x_{3,k} = \frac{1}{2}(y_{5,k} + y_{5,k-1}) + \frac{i}{2}(y_{6,k} + y_{6,k-1}) \qquad (2.45\text{c})$$

with the real dependent functions $y_{i,k}$ at the k-th discretization point. Here, their means $\frac{1}{2}(y_{i,k}+y_{i,k-1})$ between two subsequent discretization points were used, so that the following difference equations can be evaluated using information from both mesh points at $k, k-1$ [96]. We have to pay attention to the *Cauchy-Riemann* differential equations. The six real finite difference equations for the real and imaginary parts of the $x_{i,k}$ at the k-th discretization point in the "language" of the subroutine are

$$\left. \begin{array}{l} E_{1,k} = 0 = y_{1,k} - y_{1,k-1} + h/2(y_{4,k} + y_{4,k-1}) \\ E_{2,k} = 0 = y_{2,k} - y_{2,k-1} - h/2(y_{3,k} + y_{3,k-1}) \end{array} \right\} \text{Cauchy-Riemann}$$

$$E_{3,k} = 0 = y_{3,k} - y_{3,k-1} + h\,\mathfrak{Im}\big(f(\{x_{i,k}\},t)\big)$$

$$E_{4,k} = 0 = y_{4,k} - y_{4,k-1} - h\,\mathfrak{Re}\big(f(\{x_{i,k}\},t)\big)$$

$$E_{5,k} = 0 = y_{5,k} - y_{5,k-1} + h\,\mathfrak{Im}\big(g(\{x_{i,k}\},t)\big)$$

$$E_{6,k} = 0 = y_{6,k} - y_{6,k-1} - h\,\mathfrak{Re}\big(g(\{x_{i,k}\},t)\big).$$

2.3. Numerical treatment of the local equation

The $E_{i,k}$ have to vanish. For implementation, one has to calculate the 72 matrix elements

$$S_{i,j} = \frac{\partial E_{i,k}}{\partial y_{j,k-1}} \tag{2.46a}$$

$$S_{i,j+6} = \frac{\partial E_{i,k}}{\partial y_{j,k}} \tag{2.46b}$$

with $i, j \in [1, 6]$. One must thoroughly distinguish between the calculation of the derivative with respect to a real part and an imaginary part. For example, for x_1 the complex derivative forms are

$$\frac{\partial f}{\partial x_{1,k}} = 2\frac{\partial \mathfrak{Re}(f)}{\partial y_{1,k}} + 2\mathrm{i}\frac{\partial \mathfrak{Im}(f)}{\partial y_{1,k}} = 2\frac{\partial \mathfrak{Im}(f)}{\partial y_{2,k}} - 2\mathrm{i}\frac{\partial \mathfrak{Re}(f)}{\partial y_{2,k}}.$$

The factors 2 on the right hand side result from the mean between two adjacent mesh points. From this, one can see that the derivative of the real or imaginary part of a function equals the real or imaginary part of the derivative respectively. For $x_{1,k}$ this means

$$\frac{\partial \mathfrak{Re}(f)}{\partial y_{1,k}} = \frac{1}{2}\mathfrak{Re}\left(\frac{\partial f}{\partial x_{1,k}}\right) \qquad \frac{\partial \mathfrak{Re}(f)}{\partial y_{2,k}} = -\frac{1}{2}\mathfrak{Im}\left(\frac{\partial f}{\partial x_{1,k}}\right)$$

$$\frac{\partial \mathfrak{Im}(f)}{\partial y_{1,k}} = \frac{1}{2}\mathfrak{Im}\left(\frac{\partial f}{\partial x_{1,k}}\right) \qquad \frac{\partial \mathfrak{Im}(f)}{\partial y_{2,k}} = \frac{1}{2}\mathfrak{Re}\left(\frac{\partial f}{\partial x_{1,k}}\right)$$

and analogue formulas are valid for $x_{2,3}$. The formulas are used to calculate the $S_{i,j}$, $S_{i,j+6}$. E. g., for $S_{4,1}$ and $S_{4,1}$ one finds

$$S_{4,1} = \frac{\partial E_{4,k}}{\partial y_{1,k-1}} = -h\frac{\partial \mathfrak{Re}(f)}{\partial y_{1,k-1}} = -\frac{h}{2}\mathfrak{Re}\left(\frac{\partial f}{\partial x_{1,k}}\right),$$

$$S_{4,2} = \frac{\partial E_{4,k}}{\partial y_{2,k-1}} = -h\frac{\partial \mathfrak{Re}(f)}{\partial y_{2,k-1}} = \frac{h}{2}\mathfrak{Im}\left(\frac{\partial f}{\partial x_{1,k}}\right).$$

The advantage of these formulas is that the functions $f(\{x_i\}, t)$ and $g(\{x_i\}, t)$ as well as their derivatives do not need to be manually split up into their real and imaginary parts. Instead, this task is executed by the program. It allows for shorter expressions, and even if complex variables are used, f and g as well as their corresponding derivatives may be complicated enough. They are most conveniently implemented in terms of the p_i from (2.42) with $i = 1\ldots 4$, but they are not given explicitly here.

Since the integral function $g(\{x_i\}, t)$ is calculated using the trapezoidal rule, we must make some further considerations. Let the number of discretization points be M_n with $n = 1, 2$ indicating one of the two lines of integration as depicted in figure 2.2. We write the integrand of $g(\{x_i\}, t)$ as

$$\tilde{l}(\{x_{i,k}\}, \{x_{i,j}\}, t_k, t_j) = \tilde{l}_{j,k} \tag{2.47}$$

with $i = 1, 2$, $j = 1\ldots M_n$, $j \geq k$ and $k = 1\ldots M_n$ being the number of the current discretization point. The x's now have two indices: one for the number of the dependent variable (first index i) and one for the discretization point (second index k for dependence on t and second index j for dependence on t'). Then we may calculate g_k at the k-th discretization point in terms of the $\tilde{l}$ using the trapezoidal rule:

$$g_k = \frac{1}{2}\left(\tilde{l}_{k,k} + \tilde{l}_{M_n,k}\right) + \sum_{j=k+1}^{M_n-1} \tilde{l}_{j,k} . \tag{2.48}$$

When differentiating for the calculation of $S_{i,j}$, $S_{i,j+6}$, we must keep in mind that we differentiate with respect to the dependent variables $x_{i,k}$ at the current discretization point. At $j = k$ one has to differentiate $\tilde{l}_{k,k}$ with respect to any of the $x_i(t), x_i(t')$, i.e. with respect to the first two arguments of $\tilde{l}$ from equation (2.47). At $j \neq k$ one has to differentiate $\tilde{l}_{j,k}$ only with respect to $x_i(t)$, i.e. only with respect to the first argument of $\tilde{l}$ from equation (2.47). Consequently, these derivatives are evaluated in two different subroutines, one for $j = k$

2.3. Numerical treatment of the local equation

and one for $j \neq k$,

$$\frac{\partial g_k}{\partial x_{1,k}} = \frac{1}{2}\left(\frac{\partial \tilde{l}_{k,k}}{\partial x_{1,k}} + \frac{\partial \tilde{l}_{M_n,k}}{\partial x_{1,k}}\right) + \sum_{j=k+1}^{M_n-1} \frac{\partial \tilde{l}_{j,k}}{\partial x_{1,k}} \qquad (2.49a)$$

$$\frac{\partial g_k}{\partial x_{2,k}} = \frac{1}{2}\left(\frac{\partial \tilde{l}_{k,k}}{\partial x_{2,k}} + \frac{\partial \tilde{l}_{M_n,k}}{\partial x_{2,k}}\right) + \sum_{j=k+1}^{M_n-1} \frac{\partial \tilde{l}_{j,k}}{\partial x_{2,k}} \qquad (2.49b)$$

$$\frac{\partial g_k}{\partial x_{3,k}} = 0. \qquad (2.49c)$$

The boundary conditions are gained by asymptotic analysis of (2.39) for $t \to \infty$. On the first line parallel to the imaginary axis, they are applied at the lower boundary of the integration interval ($\Im\mathfrak{m}(t) \to -\infty$). At the upper boundary ($\Im\mathfrak{m}(t) \to \infty$), we demand symmetry of the real part of the solution and antisymmetry of the imaginary part of the solution. These symmetries are not restricted to the boundaries, but they must be a feature of the solution as a whole on the first line. The function $g(\{x_k\}, t)$ has the same symmetry properties as the x_i but it is not necessarily continuous at $\Im\mathfrak{m}(t) \to 0\pm$. The reason for the symmetry properties is the following: If one starts with a real initial guess on the second line on the real axis, then ϕ as well as t are real. In that case, equation (2.39) does not contain any complex coefficients. As a result, the solution on the real axis will remain real at each relaxation iteration, and the whole problem is implemented as a real problem on the real axis. At the upper integration boundary ($t \to \infty$) on the second line, we use asymptotic analysis boundary conditions again.

If P is proportional to the flow Péclet number $P_f = \rho U/D$, then it can be chosen freely but consistently with the Ivantsov condition. The corresponding values of P_c are obtained by numerical solution of the Ivantsov condition for fixed undercooling Δ using the false position method. The integral in the Ivantsov condition was treated

2. Advanced approach for dendritic growth in nonlinear systems

numerically using gaussian quadrature. This yields the full dataset (σ, P_c, P_f) equivalent to (ρ, V, U) for a certain material. Figure 3.2 in section 3.1.5 shows a representative example of the solution $\phi(t)$ parallel to the imaginary axis.

Code was written in C using *"Numerical recipes in C"* [96] and compiled with the open source *GNU C compiler* (gcc). Further details and a code demonstration are given in appendix B. For the convective systems, the program has been developed further into an integrated tool, including a command line interface, a detailed and extensive documentation as well as an installation script for the program on *Linux*-based systems.

3. Convective problems

3.1. Potential flow

This section exhibits selection of the operating state of a dendrite growing in a potential flow. The results of this section have been published in [122]. The potential flow approximation describes the idealized limit of a frictionless flow. It is applicable in superfluid helium. Thus, its range of application is very limited. However, the approximation is a good point to start with, because it provides a very simple form of the flow velocity field. It is of particular interest, how the forced flow velocity U will affect parameter selection. The problem had already been considered by Fischaleck et al. [43]. Here, the approach is extended by omitting several unnecessary linearizations. We use the model

field equations:

$$T^l_{\xi\xi} + T^l_{\eta\eta} = \psi_\eta T^l_\xi - \psi_\xi T^l_\eta \quad \text{in the liquid} \quad (3.1\text{a})$$
$$T^s_{\xi\xi} + T^s_{\eta\eta} = P_c \left(\xi T^s_\xi - \eta T^s_\eta\right) \quad \text{in the solid} \quad (3.1\text{b})$$
$$\psi_{\xi\xi} + \psi_{\eta\eta} = 0 \quad \text{potential flow} \quad (3.1\text{c})$$

interface conditions:

$$T^s = T^l \quad \text{continuity} \quad (3.2\text{a})$$
$$T^s = -\frac{1}{2}\sigma\kappa a(\theta) \quad \text{Gibbs-Thomson} \quad (3.2\text{b})$$
$$[\xi\eta_s]' = -\eta'_s \left(T^s_\xi - T^l_\xi\right) + T^s_\eta - T^l_\eta \quad \text{Stefan condition} \quad (3.2\text{c})$$

3. Convective problems

$$\psi_\xi + \eta'_s \psi_\eta = P_c(\eta_s + \eta'_s \xi) \qquad \text{mass conservation} \qquad (3.2d)$$

far field boundary conditions:

$$\lim_{\eta \to \infty} T^l = -\frac{\Delta}{P_c} \qquad \text{in the liquid} \qquad (3.3a)$$

$$\lim_{|\xi| \to \infty} T^s = 0 \qquad \text{in the solid } (\eta < 1) \qquad (3.3b)$$

$$\lim_{\eta \to \infty} \psi_\eta = \xi (P_c + P_f) \qquad \text{flow} \qquad (3.3c)$$

introduced in section 1.1. We are interested in solutions growing at a steady-state velocity $V\vec{e}_y$. The equations are written here non-dimensionally in parabolic coordinates in a moving frame of reference attached to the interface. D (contained in P_c and P_f) is assumed to be equal in both the liquid and the solid phase. Again, η_s represents the interface position. We use the stream function ψ defined in subsection 2.1, the dimensionless undercooling Δ defined in equation (1.18) and the flow Péclet number $P_f = \rho U/D$ as defined in equation (1.36). The model (3.1a)-(3.3c) is derived from (1.8a)-(1.14) in appendix A.1.1 using the non-dimensionalizations and the parabolic coordinate formulas from section 2.1. The diffusion-advection equation (1.8b) in the liquid and the Navier-Stokes equation (1.12) are invariant under *Galilean transformation*. The only field equation which is different from its form in the laboratory frame is (3.1b). The stream function does not appear in the equation describing heat diffusion in the solid. Equation (3.1c) is equivalent to the requirement $0 = \vec{\nabla} \times \vec{w} = -\Delta\psi \vec{e}_z$ of a potential flow.

From a mathematical point of view, the *Dirichlet boundary conditions* (3.2a)-(3.2b), (3.3a)-(3.3b) for the temperature and the *Neumann boundary conditions* (3.2d), (3.3c) for the stream function are sufficient to solve the problem for $T^{l,s}$ and ψ. In addition to that, the Stefan condition (3.2c) is needed, because η_s has to be calculated. Equation (3.2d) is the only interface condition applied to the stream function because in the potential flow case, there is no tangential no-slip condition due to the lack of friction in a non-viscous system. Together with (3.3c), it provides enough information to prescribe ψ, because it is a scalar function. ψ is determined up to an additive constant, because only its derivative appears in $\vec{w}$.

3.1.1. Ivantsov solution for dendritic growth in a potential flow

In the case without surface tension, we search for a temperature of the form $T = T(\eta)$, i.e. the isotherms will be parabolas. Equation (3.1a) becomes, dropping the superscript l,

$$-\psi_\xi T_\eta = T_{\eta\eta} \qquad (3.4)$$

and we conclude $\psi = \xi f(\eta)$, because ψ_ξ cannot depend on ξ regarding equation (3.4). Equation (3.1c) now reads

$$\xi f''(\eta) = 0 \quad \Rightarrow \quad f''(\eta) = 0 \quad \Rightarrow \quad f(\eta) = c_1 \eta + c_2.$$

From conditions (3.2d) and (3.3c) we find $c_1 = P_c + P_f$ and $c_2 = -P_f$ and finally

$$\psi^{\text{Iv}} = \xi \left(P_c \eta + P_f (\eta - 1) \right). \qquad (3.5)$$

The superscript "Iv" indicates the Ivantsov-like solution. Without the forced flow (represented by P_f), the calculated stream funcion would even fulfill the tangential no-slip interface condition $\vec{w} \cdot \vec{t} = -V \vec{n} \cdot \vec{e}_x$. Figure 3.1 is a cartesian field vector plot of the flow velocity $\vec{w}$ described by (3.5) (in combination with the definition (2.11)) in the vicinity of the growing solid. For $y \to \infty$, the field vectors tend to align vertically as expected regarding condition (1.14). The flow velocity is almost tangential at the interface because $\vec{w} \cdot \vec{n} \propto P_c \ll 1$.

3. Convective problems

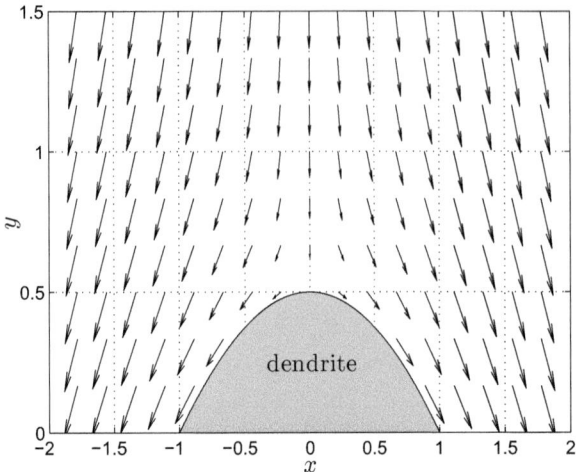

Figure 3.1.: Field vectors of the potential flow velocity $\vec{w}$ described by equation (3.5) around the growing dendrite in a cartesian frame for $\Delta = 0.01$, $P_f = 0.5$, $P_c = 0.00571$; the Ivantsov condition (3.7) is fulfilled for this parameter set.

The temperature field

$$T^{\text{Iv}}(\eta) = -e^{\frac{P_c}{2}} \int_1^\eta e^{-\frac{P_c}{2}\omega^2 - \frac{P_f}{2}(\omega-1)^2} d\omega \qquad (3.6)$$

is a solution to equation (3.4). It fulfills the interface conditions (3.2b) at zero surface tension ($\sigma = 0$) and (3.2a), because we have $T^{s,\text{Iv}} \equiv 0$ in the solid. We used the Stefan condition (3.2c) at $\eta = 1$ to determine the constant of integration:

$$1 = -\text{const.} \times e^{-\frac{P_c}{2}}.$$

3.1. Potential flow

The Ivantsov-condition for the isothermal interface reads:

$$\frac{\Delta}{P_c} = -T_\infty = \int_1^\infty e^{-\frac{P_c}{2}(\omega^2-1)-\frac{P_f}{2}(\omega-1)^2} d\omega . \qquad (3.7)$$

with the dimensionless undercooling Δ. Yet, condition (3.7) does not select a growth velocity, but it can be used to calculate P_c for given values of the undercooling Δ and P_f [66].

The equations, to which Zauderer's decomposition scheme will be applied, are set up below. We are interested in the deviation from the Ivantsov-like solution. Equations (2.13a) and (2.13b) are supplemented by

$$\psi \to \psi^{\text{Iv}} + \psi^1 \qquad (3.8)$$

with ψ^{Iv} given in (3.5) and the superscript "1" is dropped right away in the calculations below for compactness. Equations (3.1a), (3.1b) become

$$T^l_{\xi\xi} + T^l_{\eta\eta} - (\psi_\eta + \xi P_f) T^l_\xi \\ + (\psi_\xi + P_f(\eta-1)) T^l_\eta = \psi_\xi\, e^{-\frac{P_f}{2}(\eta-1)^2} \quad \text{in the liquid} \quad (3.9a)$$

$$T^s_{\xi\xi} + T^s_{\eta\eta} = 0 \qquad \text{in the solid.} \quad (3.9b)$$

Here, we have assumed the limit of small growth Péclet number ($P_c \to 0$) to apply. This is the relevant case in most experiments. It is a good approximation, if the exact solution converges continuously to the solution presented here for $P_c \to 0$. The interface conditions become

$$T^s = T^l - h \qquad \text{continuity} \qquad (3.10a)$$

$$T^s = -\frac{1}{2}\sigma\kappa a(\theta) \qquad \text{Gibbs-Thomson} \qquad (3.10b)$$

$$[\xi h]' = \left(\frac{\partial}{\partial \eta} - h'\frac{\partial}{\partial \xi}\right)(T^s - T^l) \qquad \text{Stefan condition} \qquad (3.10c)$$

3. Convective problems

$$T \to 0 \qquad \text{far field.} \qquad (3.10\text{d})$$

For derivation of the boundary conditions (3.10a)-(3.10c), we expanded $T^{\text{Iv}}\big|_{\eta=1+h}$ up to the first order in h and T was just inserted at $\eta = 1$ into (3.2a)-(3.2c), because it is assumed to be a small correction and $T_\eta h$ is already $\mathcal{O}(h^2)$,

$$T + T^{\text{Iv}}\bigg|_{\eta_s=1+h(\xi)} \approx T(\eta = 1) + T^{\text{Iv}}(\overbrace{\eta = 1}^{=0}) + \overbrace{T_\eta^{\text{Iv}}(\eta = 1)}^{=-1} h(\xi). \tag{3.11}$$

In this manner, the two-phase boundary conditions can be conveniently evaluated at $\eta = 1$, although the interface has shifted to $\eta_s = 1 + h(\xi)$. It is correct up to the first order in h. For derivation of the expanded boundary conditions for the stream function, we take the same approach:

$$\psi_{\xi,\eta}^{\text{Iv}}(\xi, 1 + h(\xi)) \approx \psi_{\xi,\eta}^{\text{Iv}}(\xi, 1) + \frac{\partial \psi_{\xi,\eta}^{\text{Iv}}}{\partial \eta}\bigg|_{\eta=1} h(\xi), \tag{3.12a}$$

$$\psi_{\xi,\eta}(\xi, 1 + h(\xi)) \approx \psi_{\xi,\eta}(\xi, 1). \tag{3.12b}$$

Equation (3.1c) is linear, so that it can also be used as the field equation for the correction of the stream function with the mass conservation condition

$$\psi_\xi + h'\psi_\eta = -P_f\,[\xi h]' \tag{3.13}$$

valid at $\eta = 1$ derived from (3.2d).

3.1.2. Continuation to the complex plane and asymptotic decomposition (potential flow)

We would like to asymptotically decompose the system (3.9a)-(3.9b) with its boundary conditions (3.10a)-(3.10d) and equation (3.1c) with

3.1. Potential flow

condition (3.13) à la Zauderer [125]. The variables $\vec{\vartheta} = (T_\xi, T_\eta)^T$ and $\vec{\vartheta}^s = (T_\xi^s, T_\eta^s)^T$ are written as linear combinations of the eigenvectors $\vec{r}_{1,2}$ (given in (2.17)) of the coefficient matrix A (given in (2.15)) of the first order systems derived from (3.9a)-(3.9b). This step was explained in section 2.2 and it is executed in equations (2.33a)-(2.33b). The decomposed equations in the liquid phase are

$$M_\xi + iM_\eta = \frac{P_f}{2}\big(\xi - i(\eta - 1)\big)M - \frac{P_f}{2}\frac{i\,[\xi h]'}{(1 + ih')}e^{-\frac{P_f}{2}(\eta-1)^2} \quad (3.14a)$$

$$N_\xi - iN_\eta = -\frac{P_f}{2}\big(\xi - i(\eta - 1)\big)M + \frac{P_f}{2}\frac{i\,[\xi h]'}{(1 + ih')}e^{-\frac{P_f}{2}(\eta-1)^2} \quad (3.14b)$$

where the solution to (3.1c) has already been inserted. These equations arise from the projection part of the Zauderer scheme. Terms of the order $\mathcal{O}(\varepsilon^2)$ were neglected in the system (3.14a)-(3.14b). Thus, the solution will only be approximate. But the simplification is necessary in order to obtain decoupled equations. Solutions to these equations can be found using characteristic coordinates:

$$s = -i(\eta - 1) \qquad \tau = \xi + i(\eta - 1) \quad (3.15a)$$
$$\bar{s} = i(\eta - 1) \qquad \bar{\tau} = \xi - i(\eta - 1)\,. \quad (3.15b)$$

The boundary conditions become

$$M = \frac{i}{2}\frac{[(1 + i\tau)\,h(\tau)]'}{(1 + ih'(\tau))} \quad (3.16a)$$

$$N = \frac{i}{2}\frac{\sigma\,[\kappa(\bar{\tau})a(\theta)]' - [(1 - i\bar{\tau})\,h(\bar{\tau})]'}{(1 - ih(\bar{\tau})')} \quad (3.16b)$$

and the solutions read:

$$M(s, \tau) = \frac{i}{2}e^{\frac{P_f}{2}(s^2 + s\tau)}\left[\frac{[(1 + i\tau)\,h]'}{(1 + ih')}\right.$$
$$\left. - \frac{2\,[\tau h]'}{\tau(1 + ih')}\left(1 - e^{-\frac{P_f}{2}s\tau}\right)\right], \quad (3.17a)$$

$$N(\bar{s},\bar{\tau}) = \frac{\mathrm{i}}{2}\left[\frac{\sigma[\kappa(\bar{\tau})a(\theta)]'}{(1-\mathrm{i}h'(\bar{\tau}))} - \frac{[(1-\mathrm{i}\bar{\tau})\,h(\bar{\tau})]'}{(1-\mathrm{i}h'(\bar{\tau}))}\right]$$
$$-\frac{P_f}{2}\bar{\tau}\int\limits_0^{\bar{s}}\left[M(-\omega,\bar{\tau}+2\omega) - \mathrm{i}\frac{[(\bar{\tau}+2\omega)\,h(\bar{\tau}+2\omega)]'}{(1+\mathrm{i}h'(\bar{\tau}+2\omega))}\mathrm{e}^{\frac{P_f}{2}\omega^2}\right]\mathrm{d}\omega\,.$$
(3.17b)

For a more detailed calculation see appendix A.1.2. A detailed discussion can also be found in [44] and [122].

3.1.3. WKB analysis of the linearized equation far from the singularity (potential flow)

We will now proceed with the asymptotic matching procedure introduced in subsection 1.2.2. Regarding the superposition (2.33a), the coefficients M and N must vanish for $\eta \to \infty$ in order to fulfill the far field boundary condition (3.10d) requiring the temperature field correction to vanish far ahead of the interface at $y \to \infty$. One can see that $M(s,\tau)$ from (3.17a) becomes zero for $s \to -\mathrm{i}\infty \Leftrightarrow \eta \to \infty$ due to the exponential function in front of the square brackets. In addition to that, we demand

$$\lim_{\bar{s}\to\mathrm{i}\infty} N(\bar{s},\bar{\tau}) = 0\,. \tag{3.18}$$

Since $\bar{\tau} \to \xi - \mathrm{i}\infty$ as $\bar{s} \to \mathrm{i}\infty$, it ensures that N behaves analytically in the far field. Considering equation (3.17b), we introduce some abbreviations and substitutions to keep formulas more convenient. To start with, we rename

$$\bar{\tau} = \xi\,.$$

Obviously, the new ξ is identical with the original parabolic coordinate only at the interface. But it can be interpreted as the analytical

3.1. Potential flow

continuation of the parabolic coordinate to the complex plane. Furthermore, we substitute

$$\omega = \frac{1}{2}(\xi' - \xi), \quad d\omega = \frac{d\xi'}{2}.$$

Setting

$$\tilde{F}(\xi) = \sigma\kappa(\xi)a(\theta) - (1 - i\xi)h(\xi)$$
$$z(\xi) = 1 + ih'(\xi) \qquad (3.19)$$
$$\bar{z}(\xi) = 1 - ih'(\xi)$$

we get

$$\tilde{F}'(\xi) = -\frac{i}{2}P_f \xi \bar{z}(\xi) \int_\xi^{i\infty} M\left(\frac{1}{2}(\xi - \xi'), \xi'\right) d\xi'$$

$$-\frac{1}{2}P_f \bar{z}(\xi) \int_\xi^{i\infty} \frac{[\xi' h]'}{z(\xi')} e^{\frac{P_f}{8}(\xi - \xi')^2} d\xi' \quad (3.20)$$

and finally using appendix A.1.3

$$\tilde{F}(\xi) = 2\int^\xi h''(\xi') \int_{\xi'}^{i\infty} M\left(\frac{1}{2}(\xi' - \xi''), \xi''\right) d\xi'' \, d\xi' + \frac{\bar{z}(\xi)}{z(\xi)}(1 + i\xi)h(\xi)$$

$$- 2i\bar{z}(\xi) \int_\xi^{i\infty} M\left(\frac{1}{2}(\xi - \xi'), \xi'\right) d\xi' + 2i\int^\xi \frac{h''(\xi')}{z^2(\xi')}(1 + i\xi')h(\xi') d\xi',$$

(3.21)

which is not quite the form we want. On the one hand, equation (3.21) exhibits a certain generality, since its form is valid for arbitrary flow approximations within the model, provided the existence of a similarity solution in the Ivantsov-like case. The proof for this universality can be found in appendix A.2.3. The idea is that the second terms in

3. Convective problems

M and N respectively always compensate each other when expressing $M\left(\frac{1}{2}(\xi-\xi'),\xi'\right)$ in terms of its own derivative with respect to ξ in (3.20). Then, integration by parts yields equation (3.21). But on the other hand, for the potential flow case we would appreciate a right hand side, that obviously vanishes for $P_f \to 0$:

$$\begin{aligned}
\sigma\kappa(\xi)a(\theta) = &(1-i\xi)\,h(\xi) + \frac{P_f}{4}e^{\frac{P_f}{8}\xi^2}\int^{\xi} e^{-\frac{P_f}{8}\xi'^2}\left[\frac{1-ih'(\xi')}{1+ih'(\xi')}\xi'\,(1-i\xi')\,h(\xi')\right.\\
&+\frac{P_f}{2}\left(1-ih'(\xi')\right)\int_{\xi'}^{i\infty}\xi''h(\xi'')\frac{\xi''-\xi'}{1+ih'(\xi'')}e^{\frac{P_f}{8}(\xi'-\xi'')^2}d\xi''\bigg]d\xi'\\
&-\frac{P_f}{2}e^{\frac{P_f}{8}\xi^2}\int^{\xi} e^{-\frac{P_f}{8}\xi'^2}\left[\left(1-ih'(\xi')\right)\int_{\xi'}^{i\infty}\xi''h(\xi'')\frac{ih''(\xi'')}{(1+ih'(\xi''))^2}e^{\frac{P_f}{8}(\xi'-\xi'')^2}d\xi''\right.\\
&+\xi'\int^{\xi'} h''(\xi'')\left(\frac{i(1+i\xi'')\,h(\xi'')}{(1+ih'(\xi''))^2}+\int_{\xi''}^{i\infty} M\left(\frac{1}{2}(\xi''-\xi'''),\xi'''\right)d\xi'''\right)d\xi''\bigg]d\xi'.
\end{aligned} \quad (3.22)$$

Equation (3.22) is the shape equation determining the correction function $h(\xi)$ for dendritic growth in a potential flow. It approximately applies in the vicinity of the singularity at $\xi = -i$. At this location in the complex plane, the curvature κ from (2.7) diverges. To counterbalance that, $h(\xi)$ on the right hand side must also behave singularly at this point. We will search for a solution $h(\xi)$ to equation (3.22) by performing asymptotic matching. I.e. we look for an approximate solution for ξ asymptotically close to the singularity. In this limit, we are going to encounter an eigenvalue equation determining σ. This will be done only in the next section. In addition to that, we look for an approximate solution far from the singularity ($|\xi+i| \gg 1$) using WKB techniques. The results from both asymptotic regions must match, if a global approximation of the solution shall be obtained.

To obtain the WKB limit, equation (3.22) is linearized in terms of h and its derivatives. On the right hand side, only the first two terms remain. The curvature from (2.7) with $|\xi| \gg 1, h$ and $h' \ll 1$ is used. We stick to the homogeneous part of the equation, since the homogeneous solution contains the exponentially decaying correction beyond all orders of a regular perturbation expansion. This correction is to be matched with the corresponding term of asymptotics beyond all orders from the inner equation. The integrand of the integral over ξ'' in the second term on the right hand side (second line in equation (3.22)) decreases strongly for $\xi'' \to i\infty$, because $h(\xi'')$, as an analytic function, must vanish for $\xi'' \to i\infty$ and $\exp\left(\frac{P_f}{8}\xi''^2\right)$ decays rapidly in this limit, more rapidly than the present factors with integer powers of ξ'' can grow. Hence in the framework of an asymptotic expansion of the integral, it is approximately proportional to the value of its integrand at the starting point $\xi'' = \xi'$. This value is zero and the term is neglected. However, it should be noted that this neglect might be a very coarse measure,

$$-\sigma\left(\frac{h''(\xi)}{\sqrt{1+\xi^2}} + \frac{\xi h'(\xi)}{(1+\xi^2)^{3/2}}\right) = (1 - i\xi)\, h(\xi)$$

$$+ \frac{P_f}{4} e^{\frac{P_f}{8}\xi^2} \int^\xi e^{-\frac{P_f}{8}\xi''^2} \xi'(1 - i\xi')\, h(\xi')\, d\xi'. \quad (3.23)$$

Equation (3.23) could have also been obtained by directly linearizing equation (3.21), integrating by parts one time and then dropping all terms, which vanish at $\xi'' = \xi'$. Equation (3.23) has the WKB solution (see appendix A.1.3)

$$h(\xi) = B_1 e^{\frac{P_f}{16}} (1 + i\xi)^{-\frac{3}{8}} (1 - i\xi)^{-\frac{5}{8}} e^{\frac{S_0(\xi)}{\sqrt{\sigma}} + \frac{P_f}{16}\xi^2} \quad (3.24)$$

with

$$S_0(\xi) = i \int_{-i}^{\xi} (1 + i\xi')^{\frac{1}{4}} (1 - i\xi')^{\frac{3}{4}}\, d\xi'. \quad (3.25)$$

3. Convective problems

This is an approximate solution far from the singularity, but it was derived from the homogeneous part of equation (3.22), which is approximately valid close to the singularity. Hence, (3.24) will only lead to the correct selection criterion, if the homogeneous solution to the exact shape equation in its linearized form contains the same transcendental corrections.

3.1.4. Transformation to a small disk around the singularity and asymptotic matching to the WKB solution (potential flow)

The next step is to convert the equation determining the function $h(\xi)$ to a local equation valid asymptotically close to the singularity at $\xi = -i$. It is reasonable to integrate (3.20) directly. With M written explicitly, that gives

$$\tilde{F}(\xi) = \frac{P_f}{4} \int_{\xi'}^{\xi} \int^{i\infty} \frac{\bar{z}(\xi')}{z(\xi'')} \left[\xi' e^{\frac{P_f}{8}(\xi'^2 - \xi''^2)} \left([(1 + i\xi'') h(\xi'')]' \right. \right.$$
$$\left. \left. - \frac{2}{\xi''} [\xi'' h(\xi'')]' \left(1 - e^{\frac{P_f}{4}\xi''(\xi'' - \xi')} \right) \right) - 2 [\xi'' h(\xi'')]' e^{\frac{P_f}{8}(\xi' - \xi'')^2} \right] d\xi'' d\xi'. \tag{3.26}$$

We apply the stretching transformation

$$\xi = -i\left(1 - \sigma^\alpha t\right) \tag{3.27a}$$
$$h(\xi) = \sigma^\alpha \phi(t) \tag{3.27b}$$

with $\alpha = \frac{2}{7}$ to (3.26). Both terms in $\tilde{F}(\xi)$ from (3.19) must have the same order of magnitude, if (3.26) shall have a solution in the flowless case $P_f = 0$. Therefore, the special choice of the scale exponent α makes sure that these terms are of the same order in the asymptotically small parameter σ. Then the local equation is solvable by finite functions. We define

$$P_1 = \frac{P_f}{4}\sigma^{\frac{2}{7}}, \quad (3.28)$$

$$b = \beta\sigma^{-\frac{4}{7}}. \quad (3.29)$$

The transformation is carried out in detail in appendix A.1.4. E. g., the transformed curvature and the transformed fourfold anisotropy function of capillary effects in $\tilde{F}$ are given by equations (A.24) and (A.23), respectively. The lowest order contribution from the flow is $\sim P_1 = \mathcal{O}(\sigma^\alpha)$, meaning that the convective effects will only be marginal unless P_f takes very large values ($P_f = \mathcal{O}(\sigma^{-\alpha})$). Let us keep terms $\propto P_1$. After having prepared all necessary ingredients, the full nonlinear eigenvalue equation determining σ is written down:

$$\phi t + \overbrace{P_1 \int_\infty^t \int^{t'} \frac{1-\dot{\phi}(t')}{1+\dot{\phi}(t'')}\left[e^{P_1(t'-t'')}\left(t''\dot{\phi} - \phi\right) + 2\dot{\phi} \times (t'-t'')\right]dt''dt'}^{\text{potential flow contribution}} = \frac{1}{\sqrt{2t+2\phi}}\underbrace{\left[\frac{\ddot{\phi}}{\left(1-\dot{\phi}^2\right)^{\frac{3}{2}}} + \frac{1+\dot{\phi}}{(2t+2\phi)\sqrt{1-\dot{\phi}^2}}\right]}_{\text{curvature}}\underbrace{\left[1 - \frac{2b\left(1-\dot{\phi}\right)^2}{(t+\phi)^2\left(1+\dot{\phi}\right)^2}\right]}_{\text{capillary anisotropy}}. \quad (3.30)$$

Equation (3.30) can be solved numerically, yielding the eigenvalue b as a function of P_1. Subsequently, one can determine σ using (3.29), provided an experimental value for β is known. At this point, there is a connection to the theory in [24]: Equation (3.30) reduces to a special form of (1.33) for $P_1 = 0$ and the eigenvalues are related by $b = \lambda^{4/7}$. The results should agree in the limit of vanishing flow (see subsection 3.1.5). The growth Péclet number P_c can then be determined numerically from equation (3.7) with $P_f = 4P_1\sigma^{-2/7}$ as input. Thus, the full dataset (σ, P_c, P_f) is obtained for a fixed value of the dimensionless undercooling Δ and the selection problem is solved

3. Convective problems

completely. For $\Delta \ll 1$, (3.7) becomes $\Delta \approx P_c\sqrt{\pi/(2P_f)}$. Together with (3.29), this yields the scaling laws

$$V \propto \Delta^{\frac{4}{3}} \beta^{\frac{7}{12}} U^{\frac{2}{3}}, \qquad (3.31a)$$

$$\rho \propto \Delta^{-\frac{2}{3}} \beta^{-\frac{7}{6}} U^{-\frac{1}{3}}. \qquad (3.31b)$$

Thus, the scaling behaviour (1.34a)-(1.34b) is changed discontinuously by the sole presence of an externally forced potential flow. σ had to be assumed independent of U for the derivation of (3.31a)-(3.31b). Equation (1.35) has to be changed:

$$V = \left(\frac{2D\Delta^4 \sigma U^2}{d_0 \pi^2}\right)^{\frac{1}{3}}. \qquad (3.32)$$

Finally, we want to show that the WKB solution (3.24) and a solution from asymptotically close to the singularity match for $t \to \infty$. For this purpose, the integral on the left hand side of (3.30) is substituted by an approximation. The right hand side of (3.30), i.e. the curvature and the anisotropy, is linearized. The leading asymptotic behaviour of ϕ for $t \to \infty$ in the resulting equation is determined by the inhomogeneity $(2t)^{-\frac{3}{2}}$. It is just powers of t. Performing asymptotic analysis this way, one can determine infinitely many orders without finding the transcendental corrections. The trick is to drop the inhomogeneity right away. Doing this, we get access to the "beyond-all-orders regime" where the matching has to be done. We find

$$\phi(t) \sim A_1 t^{-\frac{5}{8}} \exp\left(-\sqrt[4]{2}\frac{4}{7}t^{\frac{7}{4}} + \frac{P_1}{2}t - \frac{P_2}{4}t^2\right) \qquad (3.33)$$

with $P_2 = \frac{P_f}{4}\sigma^{2\alpha} = P_1 \sigma^\alpha$. This matches perfectly (see appendix A.1.4) if

$$\Im(A_1) = 0, \qquad (3.34)$$

which is a necessary requirement to any solution obtained numerically from equation (3.30). And indeed for any value of P_1, the criterion is fulfilled due to the implied numerical boundary conditions from asymptotic analysis described in section 2.3.

3.1.5. Numerical results and dependencies of the observable quantities and the selected growth mode on the potential flow

The numerical method described in section 2.3 is applied to equations (3.7) and (3.30). The real part and the imaginary part of the solution $\phi(t)$ as well as their initial guesses on the line parallel to the imaginary axis are exhibited in figure 3.2. The initial guesses ensure the (anti-)symmetry from the first iteration on to converge to the right solution. We are interested in the smallest value of b (largest growth velocity V) corresponding to the only stable solution. In the limit of vanishing flow ($P_1 \to 0$), the eigenvalue $b = 0.6122 \approx 0.42^{4/7}$ from Brener's theory [24] (see section 1.2.2) is successfully reproduced. Convergence was generally achieved in a small but experimentally relevant range of P_1. In this section, results are shown for pivalic acid (abbr. PVA, chemical formula $(CH_3)_3CCO_2H$). The important material parameters for pivalic acid are shown below.

pivalic acid:

- $\frac{L}{c} = 11.83\,K$ [115]

- $d_0 = 3.76 \cdot 10^{-3}\,\mu m = 3.76\,nm$ [101]

- $D = 7 \cdot 10^4\,\frac{\mu m^2}{s}$ [101]

- $\beta = 0.09$ [39]

- $Pr = 134.92$ [101] the *Prandtl number* defined in (3.36).

3. Convective problems

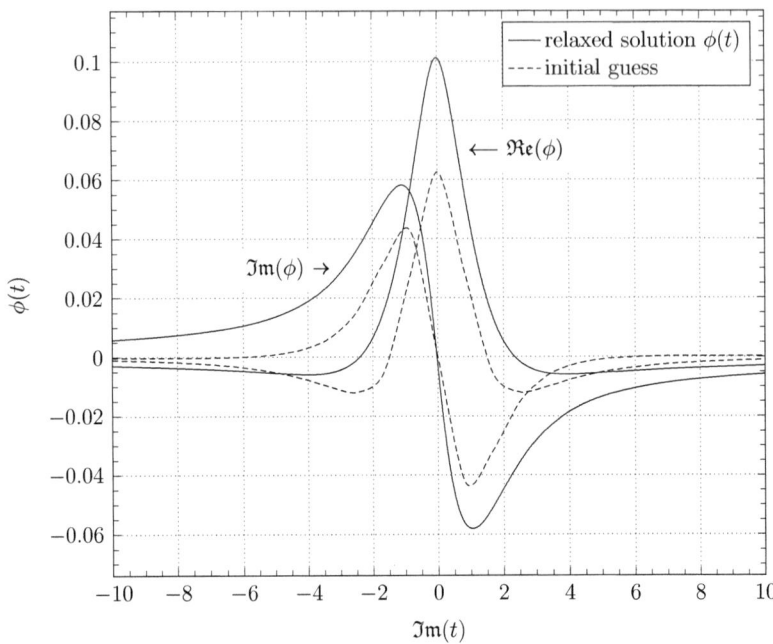

Figure 3.2.: Shape correction function $\phi(t)$ (potential flow case) in the interval of integration parallel to the imaginary axis obtained with $P_1 = 0.04$, note the symmetry properties

Pivalic acid is a transparent substance melting at $33..35\,°C$ [115]. Therefore, it is useful for experiments. It is a so called plastic crystal or organic crystal. Its properties were investigated for example in [114]. A complete table with material parameters of all the substances considered in this work can be found in appendix C.

The growth velocity V increases monotonically with increasing forced flow velocity U. This is intuitively right, because the interface is more and more flooded with undercooled liquid. For $U \to 0$, the stability parameter converges to its flowless value of $\sigma = 0.03489$ (see

3.1. Potential flow

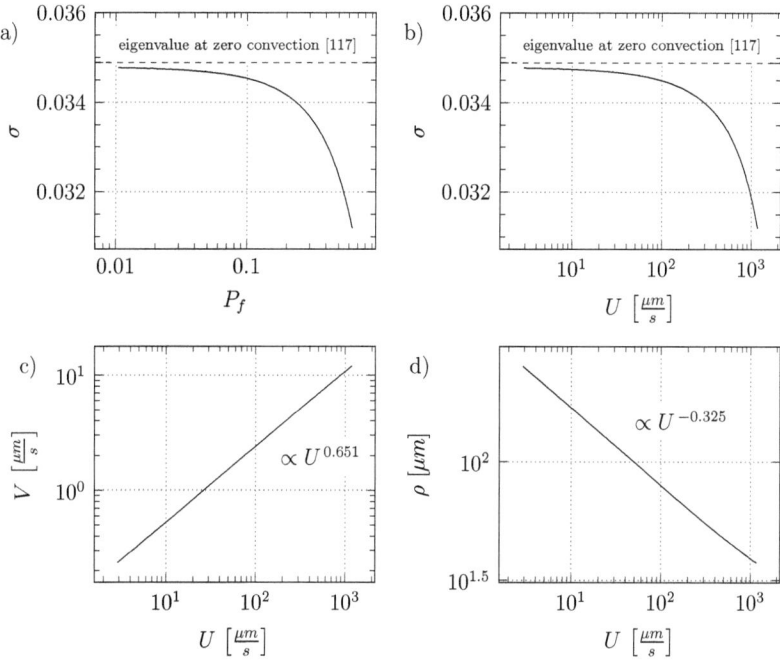

Figure 3.3.: Dependence of important growth parameters on the forced potential flow calculated numerically from equations (3.7) and (3.30) for pivalic acid ($\beta = 0.09$ [39]) with $\Delta = 0.01$ ($\widehat{=} 0.12$ K), scaling laws in c) and d) determined using least square fits to the plotted data

fig. 3.3). This value deviates somewhat from the experimental value of $\sigma = 0.022$ [52, 39]. There is a large relative uncertainty of up to 50 % in the experimental value of $\beta = 0.09$ (see section 1.3.3). This can be one reason for the deviations. We obtain $\sigma = 0.022$ by setting $\beta \approx 0.07$, which is well within the error range. Another possible reason may be that kinetic effects of atomic transfer between the phases play

3. Convective problems

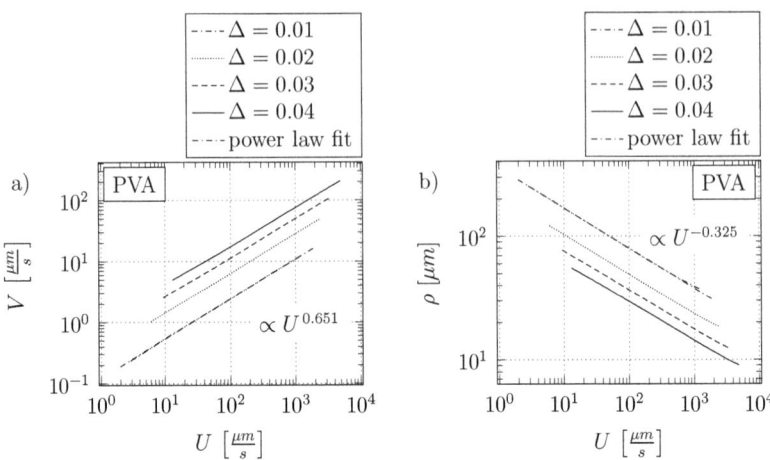

Figure 3.4.: Numerical results from selection theory in a potential flow obtained from equations (3.7) and (3.30) for pivalic acid dendrites, a) growth velocity V and b) tip curvature radius ρ as functions of the forced flow velocity U for different values of the dimensionless undercooling Δ, scaling laws determined using least square fits to the plotted data for $\Delta = 0.01$ (dash-dotted line)

an important role in pivalic acid. These effects were neglected here. However, a more quantitative analysis of this issue is given in section 4.5.

U is varied in the range $2.89 \ldots 4919.92 \, \frac{\mu m}{s}$. For V and ρ as functions of U we determine scaling laws using least square fits to the plotted data (see fig. 3.3c,d). The scaling laws are not exactly the same as the relations given in (3.31a)-(3.31b), because σ slightly depends on U in equation (3.32). In other words, b was assumed constant in (3.29) to derive (3.31a)-(3.31b). But there is a dependence of b on U because of the flow contribution in the eigenvalue equation (3.30). This dependence cannot be written down analytically and it is expected to become stronger at large P_f. Thus, the predictions in figure 3.3c,d are more accurate than the analytical approximations (3.31a)-(3.31b), especially at larger values of P_f. These scaling laws are also found at different undercoolings, see fig. 3.4. In fact, the selected eigenvalue does not depend on the undercooling here at all, because we worked in the limit $P_c \to 0$ in equation (3.30). But in reality there is a dependence [4], that cannot be neglected at larger undercoolings. Arbitrary Péclet numbers in the framework of asymptotic decomposition are investigated in section 4.4.

3.2. Oseen flow

An *Oseen flow* is more realistic than the potential flow approximation from the preceeding section 3.1. Again, no full microscopic theory is necessary to approach the problem. We use the model introduced in section 1.1 in the rest frame of the dendrite growing at constant velocity $V\vec{e}_y$ in a non-dimensional form. The field equations (3.1a) and (3.1b) for the temperature can also be used here. The flow velocity is determined by the Oseen equation

$$-(P_c + P_f)\left(\vec{e}_y \cdot \vec{\nabla}\right)\vec{w} = -\vec{\nabla}p + Pr\,\Delta\vec{w} \tag{3.35}$$

with the *Prandtl number*

$$Pr = \frac{\nu}{D} \tag{3.36}$$

and the pressure p. Equation (3.35) is a form of the Navier-Stokes equation 1.12, linearized for small flow velocities. It is an ad hoc approximation for the flow of incompressible fluids provided the *Reynolds number*

$$Re = \frac{\rho U}{\nu} \tag{3.37}$$

is small. It usually applies in highly viscous systems, where the flow velocity is generally small due to the effect of friction. Bouissou and

3. Convective problems

Pelcé used the same flow approximation in the limit $P_f \gg P_c$ [19]. In the contrary limit $P_c \gg P_f$, (3.35) is the *Stokes equation* in the frame of reference moving along with the growing dendrite. The boundary conditions (3.2a)-(3.3c) are used. In addition, the tangential no-slip condition (1.13b) is written in the moving frame of reference attached to the dendrite growing with stationary velocity (i.e. $\vec{w} \to \vec{w} + V\vec{e}_y$). This becomes $\vec{w}\cdot\vec{t} = -P_c\vec{e}_y\cdot\vec{t}$ in its dimensionless form. The parabolic coordinate forms of $\vec{t}$ and $\vec{e}_y$ from section 2.1 are inserted and $\vec{w}$ is expressed by ψ using equation (2.11):

$$\vec{w}\cdot\vec{t} = \frac{\psi_\eta - \eta'_s \psi_\xi}{\sqrt{(\xi^2 + \eta_s^2)(1 + \eta_s'^2)}} = -P_c\vec{e}_y\cdot\vec{t} = -P_c \frac{-\xi + \eta_s \eta'_s}{\sqrt{(\xi^2 + \eta_s^2)(1 + \eta_s'^2)}}.$$

It leads to the condition

$$\psi_\eta - \eta'_s \psi_\xi = P_c(\xi - \eta_s \eta'_s) \qquad (3.38)$$

valid at the interface. It ensures, that the tangential component of the flow velocity $\vec{w}$ at the two-phase boundary vanishes in the laboratory frame.

3.2.1. Ivantsov solution for dendritic growth in an Oseen flow

Neglecting capillary effects, the interface is parabolic just as all other isotherms. With the abbreviations

$$\tilde{\alpha} = \sqrt{\frac{P_c + P_f}{2Pr}} \qquad\qquad \tilde{a} = \frac{2\tilde{\alpha}\,e^{-\tilde{\alpha}^2}}{\sqrt{\pi}\,\mathrm{erfc}(\tilde{\alpha})} \qquad (3.39)$$

the flow velocity field components read

$$\psi_\xi^{\mathrm{Iv}}(\eta) = \eta\left[P_c + P_f\left(1 - \frac{\mathrm{erfc}(\tilde{\alpha}\eta)}{\mathrm{erfc}(\tilde{\alpha})}\right)\right] + \frac{P_f \tilde{a}}{2\tilde{\alpha}^2}\left[e^{\tilde{\alpha}^2(1-\eta^2)} - 1\right], \tag{3.40a}$$

$$\psi_\eta^{\text{Iv}}(\eta,\xi) = \xi\left[P_c + P_f\left(1 - \frac{\text{erfc}(\tilde{\alpha}\eta)}{\text{erfc}(\tilde{\alpha})}\right)\right]. \tag{3.40b}$$

We used the complementary error function erfc defined in (1.20). A method for derivation of (3.40a) and (3.40b) can be found in [79]. It employs a gradient form of the pressure, which fulfills a Laplace equation [43]. The calculations are shown in appendix A.2.1. The three-dimensional form of (3.40a) and (3.40b) is calculated in [10]. The complementrary error functions have to be replaced by first exponential integrals in this case. Note, that (3.40a) and (3.40b) appear also in [19], because they are solutions to the Oseen equation in the limit $P_f\tilde{a} \gg P_c$ if $\tilde{\alpha}$ is redefined by $2\tilde{\alpha}^2 = P_f/Pr$.

Let us briefly discuss this solution. Ananth and Gill stated [10], that only in three dimensions a solution of the Oseen problem can be a smooth global approximation of the flow velocity. In two dimensions, the perturbation represented by the growing parabola is too strong, much stronger than the paraboloid in three dimensions. Consider a Laplace equation: Indeed, the Green's function diverges in two dimensions in the far field. But changing the equation into diffusion type by adding the first time derivative, the Green's function becomes bounded even in two dimensions. The same argumentation can be applied to the Oseen equation (3.35). The linearized term allows for a non-divergent Green's function. The stabilizing terms in (3.40a) and (3.40b) are proportional to P_c and they represent the uniform flow arising after Galilean transformation to the rest frame of the moving interface of the dendrite. The P_f-terms are perturbative forced flow contributions to the flow velocity field, and these terms have the form of correction terms. Thus, the solution (3.40a)-(3.40b) can indeed uniformly approach the real solution of the full incompressible Navier-Stokes equations, but only in the limit $P_c \gg P_f$. Similar arguments have been put forward in [43]. However, the experimentally relevant case is usually $P_c \to 0$. Thus, any results presented in the current section are going to correspond to this very limit. Remember,

3. Convective problems

that one formal change to be made is $2\tilde{\alpha}^2 \to P_f/Pr$. So our point of view in the following will be, that we wish to solve the problem within the Oseen approximation, since only with this approximation we obtain a solid starting point, i.e. an exact similarity solution in the Ivantsov limit. The additional interesting question of how well this approximates experimental flow patterns has to be left to future investigation and to comparison with experiments.

The Ivantsov-like solution for the temperature field in the liquid is given by
$$T_\eta^{\text{Iv}}(\eta) = -\text{e}^{-I_1(\eta)} \qquad (3.41)$$
with

$$\begin{aligned} I_1(\eta) &= \int_1^\eta \psi_\xi(\omega) \, \text{d}\omega \\ &= \frac{P_c}{2}(\eta^2 - 1) + \left[\frac{P_f}{2}\eta^2 + \frac{P_f}{4\tilde{\alpha}^2}\right]\left[1 - \frac{\text{erfc}(\tilde{\alpha}\eta)}{\text{erfc}(\tilde{\alpha})}\right] \\ &\quad + \frac{P_f\tilde{\alpha}}{4\tilde{\alpha}^2}\left[1 - \eta\left(2 - \text{e}^{\tilde{\alpha}^2(1-\eta^2)}\right)\right]. \end{aligned} \qquad (3.42)$$

Solution (3.41) is derived in a manner analogous to the potential flow case. It fulfills the boundary conditions (3.2a) and (3.2c). It also fulfills the Gibbs-Thomson condition (3.2b) with $\sigma = 0$, because the temperature in the solid is just $T^{s,\text{Iv}} = 0$. The calculation of the integral $I_1(\eta)$ can be found in appendix A.2.1. We insert the temperature field (3.41) into the far field boundary condition (3.3a) and get the Ivantsov condition:

$$\Delta = P_c \int_1^\infty \text{e}^{-I_1(\omega)} \, \text{d}\omega. \qquad (3.43)$$

Figure 3.5 shows the field vectors of $\vec{w}$ in the vicinity of the dendrite and somewhat far ahead of it. The data was generated using MATLAB.

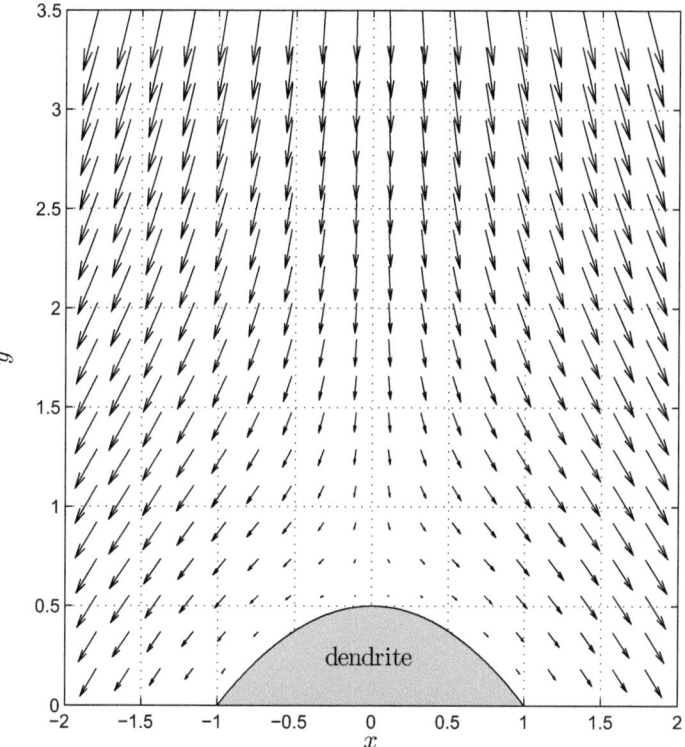

Figure 3.5.: Field vectors of the Oseen flow velocity $\vec{w}$ surrounding the growing dendrite in a cartesian moving frame of reference for succinonitrile ($Pr = 23.03$ [107]) at $\Delta = 0.02$, $P_f = 10.0$, $P_c = 0.024365$; the Ivantsov condition (3.43) is fulfilled for this dataset.

3. Convective problems

The corresponding values of P_f and P_c are a data pair resulting from the numerical solution of equation (3.43) for succinonitrile at an undercooling $\Delta = 0.02$. $P_f \tilde{a} \gg P_c$ is fulfilled. The absolute value of the vector $\vec{w}$ gradually decreases in the vicinity of the dendrite compared to the far field region. However, for $P_c > 0$ the flow does not vanish completely at the interface in the moving frame of reference. The reason is that $|\vec{w}| = P_c$ at the interface, and far ahead of it $\vec{w}$ fulfills the boundary condition $\vec{w} = -(P_f + P_c)\vec{e}_y$.

Note, that the x-component of $\vec{w}$ vanishes at the interface. $\vec{w}$ is parallel to $\vec{e}_y$ at the interface and can be decomposed into a finite normal component and a finite tangential component. This can be seen be replacing the corresponding unit vectors. Indeed our solution must have this property because, regarding the boundary conditions (3.2d) and (3.38) together, we demanded $\vec{w}(\eta = 1) = -P_c \vec{e}_y$ for the growing dendrite in the moving frame of reference.

The solutions calculated so far are now expanded about the perfectly parabolic interface. We rename the solutions (3.40a), (3.40b) and (3.41) to ψ_ξ^{Iv}, ψ_η^{Iv} and T_η^{Iv}, respectively. The upper index "Iv" is a homage to *Ivantsov*, who first calculated T^{Iv} for the flowless case [66]. In the expanded equations, P_c is set equal to zero, since this is the relevant case in most experiments. The temperature field correction is determined by

$$T^l_{\xi\xi} + T^l_{\eta\eta} - \left(\psi_\eta + \psi_\eta^{Iv}\right) T^l_\xi \\ + \left(\psi_\xi + \psi_\xi^{Iv}\right) T^l_\eta = \psi_\xi\, e^{-I_1(\eta)} \quad \text{in the liquid,} \quad (3.44a)$$

$$T^s_{\xi\xi} + T^s_{\eta\eta} = 0 \qquad \text{in the solid.} \quad (3.44b)$$

The interface is now located at $\eta_s = 1 + h(\xi)$ with an unknown function $h(\xi)$, which is to be determined from the moving boundary problem. The boundary conditions (3.10a)-(3.10d) can be used for the temperature field.

In contrast to the diffusion-advection equation (3.1a), the Oseen equation (3.35) is linear. Hence, it also holds for the correction terms and we may construct a solution to the flow velocity field using the method from [79] as before. We write its components separately:

$$\psi_\eta = -\varphi_\xi + Pr\, \chi_\xi - P_f \xi \chi, \tag{3.45a}$$
$$-\psi_\xi = -\varphi_\eta + Pr\, \chi_\eta + P_f \eta \chi. \tag{3.45b}$$

The auxiliary functions $\varphi(\xi,\eta)$ and $\chi(\xi,\eta)$ are to be calculated, but $\varphi(\xi,\eta)$ can be eliminated from the system of equations. The first equation (3.45a) is differentiated with respect to η and the second equation (3.45b) is differentiated with respect to ξ. Subsequently, the second equation is substracted from the first one.

$$\partial_\eta (3.45a) - \partial_\xi (3.45b) \Rightarrow \quad \psi_{\xi\xi} + \psi_{\eta\eta} = -P_f\left(\xi \chi_\eta + \eta \chi_\xi\right), \tag{3.46a}$$
$$\partial_\xi (3.45a) + \partial_\eta (3.45b) \Rightarrow \quad \chi_{\xi\xi} + \chi_{\eta\eta} = 2\tilde{\alpha}^2 \left(\xi \chi_\xi - \eta \chi_\eta\right). \tag{3.46b}$$

Here, $\tilde{\alpha}$ is given by $2\tilde{\alpha}^2 = P_f/Pr$. (3.46b) can be obtained by adding the derivative with respect to ξ of (3.45a) to the derivative with respect to η of (3.45b) and exploiting, that φ fulfills a Laplace equation. But its origin can also be found in appendix A.2.1, where the Ivantsov-like Oseen solution is calculated. (3.46b) is just (A.36a) written explicitly in parabolic coordinates. These equations determine the correction to the stream function. In addition to that, we have

$$\psi_\xi + h'\psi_\eta = -\xi P_f \tilde{\alpha} h h' \tag{3.47a}$$
$$\psi_\eta - h'\psi_\xi = -\xi P_f \tilde{\alpha} h \tag{3.47b}$$

as interface conditions and $\vec{w} \to 0\, (\eta \to \infty)$ as far field condition for the flow velocity field. The interface conditions are valid at $\eta = 1$. These results can be obtained by straightforward expansion of the zeroth order equations (3.2d) and (3.38) in the way demonstrated at the end of subsection 3.1.1. (3.47a)-(3.47b) are prescribed by equations (3.12a)-(3.12b). Nonlinear terms of h and its first derivative are kept in (3.47a), because these functions may diverge near the singular point $\xi = -i$ after continuing the problem to the complex plane. As

3. Convective problems

mentioned above, the relevant case is $P_f \tilde{a} \gg P_c$. Thus, terms of order $P_f \tilde{a} h$ must be kept.

3.2.2. Continuation to the complex plane and asymptotic decomposition (Oseen flow)

We look for a solution to equations (3.44a)-(3.44b) with boundary conditions (3.10a)-(3.10d) and equations (3.46a)-(3.46b) with boundary conditions (3.47a)-(3.47b). We start with the latter system describing the flow velocity. Subsequently, the result is inserted into the diffusion-advection equation (3.44a).

Can we not just find an exact analytical solution? In fact, there is a power series solution of the form

$$\chi(\xi, \eta) = \sum_{k=0}^{\infty} \sum_{l=0}^{\infty} b_k \, e_l \, \xi^k \, \eta^l \qquad (3.48)$$

to equation (3.46b) and the expansion coefficients b_k, e_l are determined by recursion relations (see equations (A.44a)-(A.44b) in appendix A.2.2). Using a separation ansatz, the *Hankel functions* and *hypergeometric functions* turn out to be solutions, too. But we have good reasons not to follow this path any further. If we did so, the boundary conditions would have to be evaluated next, if that is even possible. One would have to find a physical interpretation of the integration constants and calculating all the b_k and e_l would be much effort. Apart from that, we cannot expect to find a similar analytical solution to the diffusion-advection equation (3.44a). Thus, we look for a solution for χ, which shall be fully consistent with the approach to (3.44a).

Another possibility is to use the vorticity $\omega = \operatorname{rot} \vec{w}$. One can show, that $\operatorname{rot} \vec{w} = -\Delta \psi \vec{e}_z$. Furthermore, if we apply $\vec{\nabla} \times$ to the Oseen equation (3.35), we get

$$\Delta \psi + \omega = 0 \tag{3.49a}$$
$$\Delta \omega + 2\tilde{\alpha}^2 \left(\eta \omega_\eta - \xi \omega_\xi\right) = 0 \tag{3.49b}$$

as a substitute for the system (3.46a)-(3.46b). Both ways are pretty elegant forms of the Oseen equation. (3.46b) and (3.49b) are actually the same equation expressed by different quantities. But by introducing the vorticity, the information contained in the pressure is lost. It has to be regained by an additional integration, which does not have to be performed when using the system (3.46a)-(3.46b). Of course, one has to deal with the pressure in one or the other way. But (3.46a) contains only derivatives of χ. This is a crucial advantage. $\chi_{\xi,\eta}$ obtained from (3.46b) do not have to be integrated, as we would have to do with $\omega_{\xi,\eta}$ before inserting into (3.49a).

The application of the asymptotic decomposition procedure à la Zauderer as described in section 2.2 and the solution of the decomposed equations constitute quite technical undertakings in this case. The calculations are relegated to appendix A.2.2. We just write down the solutions here, which are used for asymptotic matching. The characteristic coordinates given in (3.15a) and (3.15b) also apply here. Let again $\vec{r}_{1,2}$ be the eigenvectors of the coefficient matrix A of the first order systems derived from (3.46a) and (3.44a). These vectors are given in (2.17), and A is defined in (2.15). With the superposition $\vec{v} = (\psi_\xi, \psi_\eta)^T = \beta^{(1)} \vec{r}_1 + \varepsilon \beta^{(2)} \vec{r}_2$, the functions

$$\beta^{(1)}(s,\tau) = -\frac{1}{2} P_f \tau \tilde{a} h(\tau) e^{\tilde{\alpha}^2 \left[s^2 - is(1+i\tau)\right]} \tag{3.50a}$$

$$\beta^{(2)}(\bar{s}, \bar{\tau}) = -\frac{i}{2} P_f \tilde{\alpha}^2 \tilde{a} \left(1 + i\bar{\tau}\right) \int_0^{\bar{s}} \left(\bar{\tau} + 2\omega\right) h(\bar{\tau} + 2\omega) \\ \times e^{\tilde{\alpha}^2 \left[-\omega^2 + i\omega(1+i\bar{\tau})\right]} \, d\omega - \frac{1}{2} P_f \bar{\tau} \tilde{a} h(\bar{\tau}) \tag{3.50b}$$

3. Convective problems

are determined. The solution $\vec{v}$ and the superposition $\vec{\vartheta} = (T_\xi, T_\eta)^{\mathrm{T}} = M\vec{r}_1 + \varepsilon N\vec{r}_2$ are inserted into the first order system derived from (3.44a). Using the abbreviations

$$a_{1,2}(s,\tau) = \frac{1}{2}\left[\psi_\eta^{\mathrm{Iv}}(s+\tau, 1+\mathrm{i}s) \mp \mathrm{i}\psi_\xi^{\mathrm{Iv}}(1+\mathrm{i}s)\right] \tag{3.51}$$

the decomposed equations for the coefficients M and N can be written down in their respective characteristic coordinates:

$$M_s = a_1(s,\tau)M + \frac{1}{2}\beta^{(1)}(s,\tau)\mathrm{e}^{-I_1(1+\mathrm{i}s)}, \tag{3.52a}$$

$$N_{\bar{s}} = -a_1(-\bar{s}, \bar{\tau}+2\bar{s})M - \frac{1}{2}\beta^{(1)}(-\bar{s}, \bar{\tau}+2\bar{s})\mathrm{e}^{-I_1(1-\mathrm{i}\bar{s})}. \tag{3.52b}$$

These equations already contain the asymptotically consistent solution to the system (3.46a)-(3.46b). $\psi_{\xi,\eta}$ do not appear in $a_{1,2}$, because $\beta^{(1)}$ vanishes when comparing $\mathrm{i}\psi_\xi$ and ψ_η. The $\beta^{(2)}$-term was dropped in (3.52a)-(3.52b), since it remains finite at the singularity ($\xi = -\mathrm{i}$). I.e. the $\beta^{(2)}$-term is $\mathcal{O}(\varepsilon^2)$ within the Zauderer scheme in the first order system corresponding to equation (3.44a). The solutions to (3.52a) and (3.52b) are

$$M(s,\tau) = \overbrace{\exp\left[\int_0^s a_1(\omega, \tau)\,\mathrm{d}\omega\right]}^{\text{homogenous solution}} \overbrace{\left[\frac{\mathrm{i}\left[(1+\mathrm{i}\tau)\,h(\tau)\right]'}{2\left(1+\mathrm{i}h'(\tau)\right)} - \frac{P_f}{4}\tilde{a}\tau h(\tau)\right.}^{\text{integration constant}}$$

$$\left. \times \int_0^s \exp\left(\tilde{a}^2\left(\omega^2 - \mathrm{i}\omega\left(1+\mathrm{i}\tau\right)\right) - \int_0^\omega a_2(\omega', \tau)\,\mathrm{d}\omega'\right)\mathrm{d}\omega\right]$$

$$\tag{3.53a}$$

$$N(\bar{s},\bar{\tau}) = -\int_0^{\bar{s}} a_1(-\omega, \bar{\tau}+2\omega) M(-\omega, \bar{\tau}+2\omega) \,\mathrm{d}\omega$$

$$+ \frac{P_f}{4}\tilde{a} \int_0^{\bar{s}} (\bar{\tau}+2\omega) h(\bar{\tau}+2\omega) e^{\tilde{a}^2\left[-\omega^2 + \mathrm{i}\omega(1+\mathrm{i}\bar{\tau})\right] - I_1(1-\mathrm{i}\omega)} \,\mathrm{d}\omega$$

$$+ \frac{\mathrm{i}}{2}\left[\frac{\sigma[\kappa(\bar{\tau})a(\theta)]'}{1-\mathrm{i}h'(\bar{\tau})} - \frac{[(1-\mathrm{i}\bar{\tau})h(\bar{\tau})]'}{1-\mathrm{i}h'(\bar{\tau})}\right]$$

(3.53b)

where the boundary conditions (3.16a) and (3.16b) have been applied.

3.2.3. WKB-analysis of the linearized equation far from the singularity (Oseen flow)

As in the beginning of subsection 3.1.3, we require $N(\bar{s} \to \mathrm{i}\infty, \bar{\tau}) \to 0$ to fulfill the far field boundary condition (3.10d), and we set $\bar{\tau} \to \xi$ and $\omega = \frac{1}{2}(\xi' - \xi)$. Doing this yields an equation determining the function $h(\xi)$:

$$\sigma[\kappa(\xi)a(\theta)]' = [(1-\mathrm{i}\xi)h(\xi)]' + \mathrm{i}(1-\mathrm{i}h'(\xi)) \times \Bigg[$$
$$\int_{\mathrm{i}\infty}^{\xi} a_1\left(\frac{1}{2}(\xi-\xi'), \xi'\right) M\left(\frac{1}{2}(\xi-\xi'), \xi'\right) \mathrm{d}\xi' \qquad (3.54)$$
$$- \frac{P_f}{4}\tilde{a} \int_{\mathrm{i}\infty}^{\xi} \xi' h(\xi') e^{\tilde{a}^2\left[\frac{1}{4}(\xi^2-\xi'^2) - \frac{\mathrm{i}}{2}(\xi-\xi')\right] - I_1\left(1+\frac{\mathrm{i}}{2}(\xi-\xi')\right)} \mathrm{d}\xi' \Bigg].$$

One could get here without calculating a special form of $\psi_{\xi,\eta}$ or $\psi_{\xi,\eta}^{\mathrm{Iv}}$. I.e. if one writes a_1 and I_1 in terms of $\psi_{\xi,\eta}$ (which we did) and calculates M the same way as above without inserting a special form of $\psi_{\xi,\eta}$ or $\psi_{\xi,\eta}^{\mathrm{Iv}}$, the shape equation (3.54) holds for arbitrary flows, if an

3. Convective problems

Ivantsov-like solution exists. Unfortunately, it is difficult to analyze and it is a hopeless matter trying to solve it exactly. Instead, we perform asymptotic matching again. In this subsection, we look for an approximate solution of the linearized form of (3.54) far from the singularity ($|\xi + \mathrm{i}| \to \infty$). Subsequently, the equation is transformed close to the singularity, and the resulting eigenvalue problem is solved numerically.

We start with equation (3.21) because it is also valid here, although it is derived in the section about the potential flow approximation. The proof for its general applicability can be found at the beginning of appendix A.2.3. Linearizing the Oseen-flow version of (3.21) and using the abbreviation

$$I_3(-\bar{s}', \xi') = \int_0^{\frac{1}{2}(\xi - \xi')} a_1(\omega, \xi') \, \mathrm{d}\omega \qquad (3.55)$$

we find

$$-\sigma \left(\frac{h''(\xi)}{\sqrt{1+\xi^2}} + \frac{\xi h'(\xi)}{(1+\xi^2)^{3/2}} \right) = 2h(\xi)$$

$$+ \frac{\mathrm{i}}{2} P_f \tilde{a} \int_\xi^{\mathrm{i}\infty} \xi' h(\xi') \, \mathrm{e}^{I_3(-\bar{s}', \xi')} \int_0^{\frac{1}{2}(\xi - \xi')} \mathrm{e}^{\tilde{a}^2 \left(\omega^2 - \mathrm{i}\omega(1 + \mathrm{i}\xi') \right)}$$

$$\times \exp \left(- \int_0^\omega a_2(\omega', \xi') \, \mathrm{d}\omega' \right) \mathrm{d}\omega \, \mathrm{d}\xi' + \int_\xi^{\mathrm{i}\infty} [(1 + \mathrm{i}\xi') \, h(\xi')]' \, \mathrm{e}^{I_3(-\bar{s}', \xi')} \, \mathrm{d}\xi' .$$

(3.56)

The first integral on the right hand side is integrated by parts. This is reasonable, because one factor appears as a derivative. The calculations are shown in appendix A.2.3. The explicit calculation of I_3 can also be found in the designated appendix part. We get the following form of the shape equation for WKB-analysis:

$$-\sigma\left(\frac{h''(\xi)}{\sqrt{1+\xi^2}} + \frac{\xi h'(\xi)}{(1+\xi^2)^{3/2}}\right) = (1-\mathrm{i}\xi)\,h(\xi) + \frac{\mathrm{i}}{4}P_f\tilde{a}\int_{\mathrm{i}\infty}^{\xi} h(\xi')\,\mathrm{e}^{I_3(-\tilde{s}',\xi')} \times \Bigg[$$
$$\frac{1+\mathrm{i}\xi'}{Re}\left(1 - \mathrm{e}^{\tilde{a}^2\left(\frac{1}{4}(\xi-\xi')^2 - \frac{1}{2}(\xi-\xi')\right)}\right) - \frac{(1+\xi'^2)}{\tilde{a}}\left(1 - \frac{\mathrm{erfc}\left(\tilde{a}\left(1 + \frac{\mathrm{i}}{2}(\xi-\xi')\right)\right)}{\mathrm{erfc}(\tilde{a})}\right)$$
$$-2\xi'\int_0^{\frac{1}{2}(\xi-\xi')} \mathrm{e}^{\tilde{a}^2\left(\omega^2 - \mathrm{i}\omega(1+\mathrm{i}\xi')\right)}\exp\left(-\int_0^{\omega} a_2(\omega',\xi')\,\mathrm{d}\omega'\right)\mathrm{d}\omega\Bigg]\,\mathrm{d}\xi'.$$

(3.57)

Unfortunately, this linear equation (3.57), which was derived from (3.56), is not simple enough for WKB analysis. The integrals constitute a problem, and they cannot be eliminated by differentiating the whole equation once with respect to ξ, because ξ also appears in the integrand. For this reason, the P_f-contributions to the WKB solution are not calculated in this section:

$$-\sigma\left(\frac{h''(\xi)}{\sqrt{1+\xi^2}} + \frac{\xi h'(\xi)}{(1+\xi^2)^{3/2}}\right) = (1-\mathrm{i}\xi)\,h(\xi). \qquad (3.58)$$

In (3.58), a linearized form of the curvature $\kappa(\xi)$ has been used. The inhomogeneity was dropped to access the transcendental corrections, which are contained only in the homogeneous solution. We find the WKB solution

$$h(\xi) = B_2\,(1+\mathrm{i}\xi)^{-\frac{3}{8}}(1-\mathrm{i}\xi)^{-\frac{5}{8}}\,\mathrm{e}^{\frac{S_0(\xi)}{\sqrt{\sigma}}}. \qquad (3.59)$$

to (3.58), again using A.2.3. S_0 was given in (3.25). (3.59) is the "outer solution". B_2 is an integration constant. Some terms were integrated with lower boundary $-\mathrm{i}$, i.e. starting with the singular point. But some terms were integrated indefinitely, because they diverge at $\xi = -\mathrm{i}$. This is not surprising, since the WKB theory is valid only far away from this point. In the vicinity of this point, the WKB solution breaks down and we have to construct another solution.

3.2.4. Transformation to a small disk around the singularity and asymptotic matching to the WKB solution (Oseen flow)

The WKB solution (3.59) breaks down in the vicinity of $\xi = -\mathrm{i}$. To construct a smooth global approximation of the function $h(\xi)$ in the complex plane, a better description close to the singularity needs to be found. For this purpose, (3.54) is integrated directly:

$$\sigma\kappa(\xi) = (1 - \mathrm{i}\xi)\,h(\xi) + \mathrm{i}\int_{\mathrm{i}\infty}^{\xi}\int^{\xi} (1 - \mathrm{i}h'(\xi')) \times \Bigg[$$
$$a_1\left(\frac{1}{2}(\xi' - \xi''), \xi''\right) M\left(\frac{1}{2}(\xi' - \xi''), \xi''\right)$$
$$\underbrace{-\frac{P_f}{4}\tilde{a}\xi''h(\xi'')\,\mathrm{e}^{\tilde{a}^2\left[\frac{1}{4}(\xi'^2 - \xi''^2) - \frac{1}{2}(\xi' - \xi'')\right] - I_1\left(1 + \frac{1}{2}(\xi' - \xi'')\right)}}_{=t_6}\Bigg]\,\mathrm{d}\xi''\,\mathrm{d}\xi'.$$

(3.60)

This equation is transformed onto a small disc around the singularity via the transformation (3.27a)-(3.27b). The analogue stretching transformation is used for the primed and double-primed quantities. The first terms on both sides of (3.60) are only of the same order of σ, if $\alpha = \frac{2}{7}$ as in the potential flow case. α is not to be mixed up with the abbreviation $\tilde{\alpha}$. We find $\kappa = \mathcal{O}\left(\sigma^{\left(-\frac{3}{7}\right)}\right)$. The explicit forms of the curvature contribution and the anisotropy factor on the new scale can be read off equation (3.30). The transformation of the double integral in (3.60) is a bigger challenge. This part is completely relegated to appendix A.2.4. We use the definition (3.29) and define

$$P_3 = \frac{P_f}{4}\tilde{a}\sigma^\alpha, \qquad (3.61)$$

meaning $P_3 = P_1\tilde{a}$, and write the transformed equation:

$$\frac{1}{\sqrt{2t+2\phi}}\left[\frac{\ddot{\phi}}{\left(1-\dot{\phi}^2\right)^{3/2}}+\frac{1+\dot{\phi}}{(2t+2\phi)\sqrt{1-\dot{\phi}^2}}\right]\left[1-\frac{2b(1-\dot{\phi})^2}{(t+\phi)^2(1+\dot{\phi})^2}\right]-t\phi$$
$$=P_3\int_\infty^t\int^{t'}\frac{1-\dot{\phi}(t')}{1+\dot{\phi}(t'')}\left[\dot{\phi}(t'')(t'-t'')+\left(1+\dot{\phi}(t'')\right)\phi(t'')\right]\mathrm{d}t''\mathrm{d}t'.$$
(3.62)

Brener showed, that at arbitrary growth Péclet number P_c, a corresponding contribution of the order σ^α arises on the right hand side of equation (3.62) [24]. The convective P_3-term is of the same order of magnitude, which makes sense, because in the moving frame of reference the P_c-contributions act as a uniform flow. See also section 4.4 for arbitrary growth Péclet numbers P_c in the framework of asymptotic decomposition. Thus, equation (3.62) is valid in the limit $P_f\tilde{a}\gg P_c$.

In the limit $t\to\infty$, the solution to (3.62) should match the WKB-solution (3.59). Using appendix A.2.4, we find the asymptotic approximation

$$\phi(t)=A_1t^{-\frac{5}{8}}\exp\left(-\sqrt[4]{2}\frac{4}{7}t^{\frac{7}{4}}\right),\tag{3.63}$$

fulfilling the physical condition of a smooth tip only if

$$0=\mathfrak{Re}\left(\left.\frac{\mathrm{d}\eta_s}{\mathrm{d}\xi}\right|_{\xi=0}\right)\propto\mathfrak{Im}\left(A_1\right).\tag{3.64}$$

A zero imaginary part of the numerical constant A_1 is a necessary requirement to any solution obtained from equation (3.62).

3.2.5. Scaling laws in the large flow Péclet number limit

The approximate potential flow scaling laws (3.31a)-(3.31b) hold for $\Delta\ll 1$. They are not exact, because the dependence of σ on U was

3. Convective problems

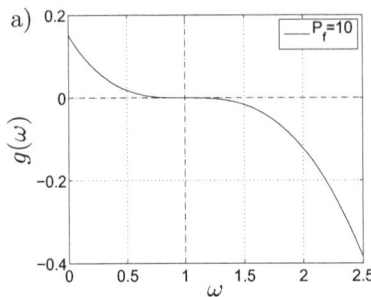 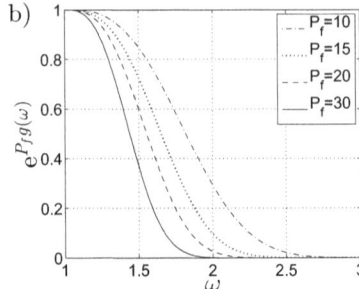

Figure 3.6.: The function a) $g(\omega) = -I_1(\omega)/P_f$ with $I_1(\omega)$ from (3.42) for $P_c = 0$, $P_f = 10$ and $Pr = 23.03$ (SCN), b) integrand $e^{P_f g(\omega)}$ decaying more rapidly for larger values of P_f

neglected in their derivation. A general analogon in the Oseen flow case cannot be given, because the Ivantsov condition (3.43) does not simplify enough for $P_c \to 0$. But scaling laws can be found in the limit $P_f \gg 1$. Consider the exponent of the integrand in (3.43). If we write

$$g(\omega) := \lim_{P_c \to 0} \left(-\frac{I_1(\omega)}{P_f}\right) \qquad (3.65)$$

with $I_1(\omega)$ from (3.42), then the function $g(\omega)$ has a saddle point at $\omega = 1$. For $P_f \gg 1$, the integrand $e^{P_f g(\omega)}$ decays rapidly for $\omega > 1$ (see figure 3.6). Thus, the integral is dominated by its starting point (the maximum at $\omega = 1$). It can be expanded using *Laplace's method*, i.e. the exponent is expanded in a Taylor series around $\omega = 1$ up to the first non-vanishing order. Since it is a saddle point, the first non-vanishing contribution will be third order. We write down the derivatives of $g(\omega)$:

$$\frac{dg}{d\omega} = \lim_{P_c \to 0} \left(-\frac{\psi_\varepsilon(\omega)}{P_f}\right) \qquad (3.66a)$$

$$\frac{d^2 g}{d\omega^2} = -\left[1 - \frac{\operatorname{erfc}(\tilde{\alpha}\omega)}{\operatorname{erfc}(\tilde{\alpha})}\right] \qquad (3.66b)$$

3.2. Oseen flow

$$\frac{d^3 g}{d\omega^3} = -\tilde{a}\, e^{\tilde{\alpha}^2 (1-\omega^2)} . \tag{3.66c}$$

One finds $g'(1) = 0$, $g''(1) = 0$ and $g'''(1) = -\tilde{a}$ and thus

$$g(\omega)\Big|_{\omega=1} \approx -\frac{\tilde{a}}{6}(\omega - 1)^3 . \tag{3.67}$$

The integral is approximated as

$$\int_1^\infty e^{-I_1(\omega)}\, d\omega \approx \int_1^\infty e^{-\frac{P_f \tilde{a}}{6}(\omega-1)^3}\, d\omega = \frac{1}{3}\Gamma\left(\frac{1}{3}\right)\sqrt[3]{\frac{6}{\tilde{a}}} P_c P_f^{-\frac{1}{3}} \tag{3.68}$$

where it was written in terms of the gamma function. For large P_f, one has $\tilde{\alpha} = \sqrt{P_f/(2Pr)} \gg 1$. Thus, the asymptotic behaviour

$$\text{erfc}(\tilde{\alpha}) \sim \frac{e^{-\tilde{\alpha}^2}}{\sqrt{\pi}\tilde{\alpha}} \qquad \tilde{\alpha} \to \infty \tag{3.69}$$

of the complementary error function can be inserted into $\tilde{a}$ given in (3.39) in order to derive scaling laws from (3.68). It leads to $\tilde{a} \sim 2\tilde{\alpha}^2 = P_f/Pr$ and the Ivantsov condition becomes

$$\frac{\Delta}{P_c} = \frac{1}{3}\Gamma\left(\frac{1}{3}\right)\sqrt[3]{6Pr}\, P_f^{-\frac{2}{3}} . \tag{3.70}$$

This, together with $\sigma \propto \beta^{7/4}$ from the preceeding section, yields the scaling laws

$$V \propto \Delta^{\frac{6}{5}} \beta^{\frac{7}{20}} U^{\frac{4}{5}} , \tag{3.71a}$$

$$\rho \propto \Delta^{-\frac{3}{5}} \beta^{-\frac{21}{20}} U^{-\frac{2}{5}} . \tag{3.71b}$$

These laws represent a good description in an intermediate asymptotic range of P_f. However, at very large P_f equation (3.70) is not the best approximation of the Oseen Ivantsov condition (3.43). The reason

3. Convective problems

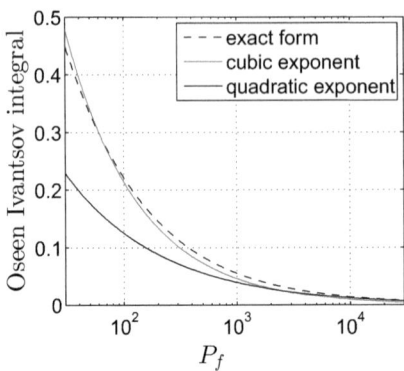

Figure 3.7.: Approximations of the Oseen Ivantsov integral from (3.43) at large flow Péclet number P_f for succinonitrile ($Pr = 23.03$), dashed graph: exact form (left hand side of (3.68)), grey graph: cubic approximation (right hand side of (3.70)), black graph: quadratic approximation (right hand side of (3.72))

is, that for $\tilde{\alpha} \gg 1$, the absolute value of the second derivative $g''(\omega)$ given in (3.66b) increases very rapidly from 0 to 1 as ω becomes slightly larger than 1. Therefore, a quadratic exponent yields better agreement of the approximated and the real integrand for $\omega > 1$ at $\tilde{\alpha} \gg 1$. It can be seen analytically by inserting the asymptotic behaviour (3.69) of the complementary error function into $g(\omega)$:

$$g(\omega) = -\frac{1}{2}\left(\omega^2 + \frac{1}{2\tilde{\alpha}^2}\right)\left(1 - \frac{\text{erfc}(\tilde{\alpha}\omega)}{\text{erfc}(\tilde{\alpha})}\right)$$
$$-\frac{1}{2}\frac{\tilde{\alpha}}{2\tilde{\alpha}^2}\left[1 - \omega\left(2 - e^{\tilde{\alpha}^2(1-\omega^2)}\right)\right]$$
$$\sim -\frac{1}{2}\omega^2\left[1 - \frac{1}{\omega}e^{\tilde{\alpha}^2(1-\omega^2)}\right] - \frac{1}{2}\left[1 - \omega\left(2 - e^{\tilde{\alpha}^2(1-\omega^2)}\right)\right] \quad \tilde{\alpha} \to \infty$$
$$= -\frac{1}{2}(\omega - 1)^2 + \frac{1}{2}e^{\tilde{\alpha}^2(1-\omega^2)}(\omega - \omega) = -\frac{1}{2}(\omega - 1)^2.$$

The resulting integral is Gaussian and can be solved analytically:

$$\frac{\Delta}{P_c} = \int_1^\infty e^{-\frac{P_f}{2}(\omega-1)^2} \, d\omega = \sqrt{\frac{\pi}{2P_f}}. \qquad (3.72)$$

This yields the same scaling laws (3.31a)-(3.31b) and the same explicit relation (3.32) for the growth velocity V as in the potential flow case. But here they are restricted to the limit $P_f \gg 1$.

Figure 3.7 shows the different forms of the Ivantsov integrals as functions of P_f for succinonitrile ($Pr = 23.03$). At $P_f \approx 62$, the exact integral from the left hand side of (3.68) (dashed graph) and its cubic approximation given on the right hand side of (3.70) (grey graph) take the same value. Here, the corresponding integrands $\exp\left(-I_1(\omega)\right)$ and $\exp\left(-\frac{P_f \tilde{a}}{6}(\omega-1)^3\right)$ are almost the same function on the whole range of integration. But beyond that point, the deviation between the exact integral and its cubic approximation increases. The solid black graph is a plot of the right hand side of (3.72), where the quadratic approximation of the exponent is used. It approaches the exact form slowly but monotonically. For $P_f \gtrsim 2500$, the solid black graph has crossed the grey graph and the deviation between exact and quadratic is smaller than the deviation between exact and cubic.

In three dimensions, the stream function can be calculated using the paraboloid coordinates $x = \eta\xi \cos\varphi$, $y = \eta\xi \sin\varphi$ and $z = \frac{1}{2}(\eta^2 - \xi^2)$. Its derivatives read

$$\psi_\xi^{3D}(\xi,\eta) = \xi\eta^2 \left[P_c + P_f \left(1 - \frac{E_1\left(\tilde{\alpha}^2 \eta^2\right)}{E_1\left(\tilde{\alpha}^2\right)}\right)\right] \\ - \frac{P_f \xi e^{-\tilde{\alpha}^2}}{\eta \tilde{\alpha}^2 E_1\left(\tilde{\alpha}^2\right)} \left(1 - e^{\tilde{\alpha}^2\left(1-\eta^2\right)}\right) \qquad (3.73a)$$

$$\psi_\eta^{3D}(\xi,\eta) = \xi^2\eta \left[P_c + P_f \left(1 - \frac{E_1\left(\tilde{\alpha}^2 \eta^2\right)}{E_1\left(\tilde{\alpha}^2\right)}\right)\right] \qquad (3.73b)$$

if surface tension is neglected. Here, E_1 is the first exponential integral:

$$E_1(x) = \int_x^\infty \frac{e^{-t}}{t} dt \ . \tag{3.74}$$

These forms were also given by Ananth and Gill [10] with $\tilde{\alpha}^2 = P_f/(2Pr)$. The corresponding Ivantsov condition is

$$\frac{\Delta}{P_c} = \int_1^\infty \exp\left(-\int_1^\omega \frac{\psi_\xi^{3D}(\xi,\omega')}{\xi\omega'} d\omega'\right) \frac{d\omega}{\omega} \tag{3.75}$$

and thus we define the function

$$\begin{aligned}g^{3D}(\omega) &= \lim_{P_c \to 0}\left(-\int_1^\omega \frac{\psi_\xi^{3D}(\xi,\omega')}{P_f \xi \omega'} d\omega'\right) \\ &= -\frac{1}{2}\left(\omega^2 + \frac{1}{\tilde{\alpha}^2}\right)\left(1 - \frac{E_1\left(\tilde{\alpha}^2 \omega^2\right)}{E_1\left(\tilde{\alpha}^2\right)}\right) \\ &\quad + \frac{e^{-\tilde{\alpha}^2}}{\tilde{\alpha}^2 E_1\left(\tilde{\alpha}^2\right)}\left[\ln\omega + \frac{1}{2}\left(1 - e^{\tilde{\alpha}^2(1-\omega^2)}\right)\right]\end{aligned} \tag{3.76}$$

to be used in the exponent in 3D. Inserting the asymptotic behaviour

$$E_1\left(\tilde{\alpha}^2\right) \sim \frac{e^{-\tilde{\alpha}^2}}{\tilde{\alpha}^2} \qquad \tilde{\alpha}^2 \to \infty \tag{3.77}$$

into (3.76), we find the quadratic approximation

$$\frac{\Delta}{P_c} = \int_1^\infty e^{-P_f(\omega-1)^2} \frac{d\omega}{\omega} \tag{3.78}$$

where the logarithm was expanded about $\omega = 1$ up to the quadratic order. This integral cannot be solved analytically due to the factor $\frac{1}{\omega}$ in the integrand.

3.2.6. Numerical results and dependencies of the observables quantities and the selected growth mode on the Oseen flow

The set (3.43), (3.62) is solved numerically using the method described in section 2.3. (3.62) applies in the limit $P_f \tilde{a} \gg P_c$ and it turned out, that $\tilde{a}$ is less than unity for all materials considered in this subsection and for all values of P_f in use. Consequently, $P_f \gg P_c$ definitely holds, if $P_f \tilde{a} \gg P_c$ is fulfilled. It allows us to set $\tilde{\alpha}^2 \approx P_f/(2Pr)$ in the Ivantsov condition as well as in the local equation. This conveniently shortens the numerical working flow, because if (3.62) is completely free of P_c, the resulting dataset (P_3, b) is independent of the undercooling and it has to be determined only once. Subsequently, it is used as input for (3.43) to calculate P_c at different values of Δ. Results were obtained among others for succinonitrile (abbr. SCN, chemical formula $C_2H_4(CN)_2$), a substance with two nitrile groups.

succinonitrile:

- $\frac{L}{c} = 23.12$ [102]

- $d_0 = 2.79\,nm$ [101], [102] (thermal capillary length)

- $D = 1.13 \cdot 10^5\,\frac{\mu m^2}{s}$ [101], [102] (thermal diffusivity)

- $\beta = 0.075$ [89], [101], [102]

- $Pr = 23.03$ [107]

- $\gamma = 8.9 \cdot 10^{-3}\,\frac{J}{m^2}$ [102], [120]

- $\nu = 26 \cdot 10^5\,\frac{\mu m^2}{s}$ [120].

In the flowless case ($P_3 = 0$), we reproduce the eigenvalue $b = 0.6122$ of Brener [24] and Tanveer [117]. For succinonitrile with $\beta = 0.075$,

3. Convective problems

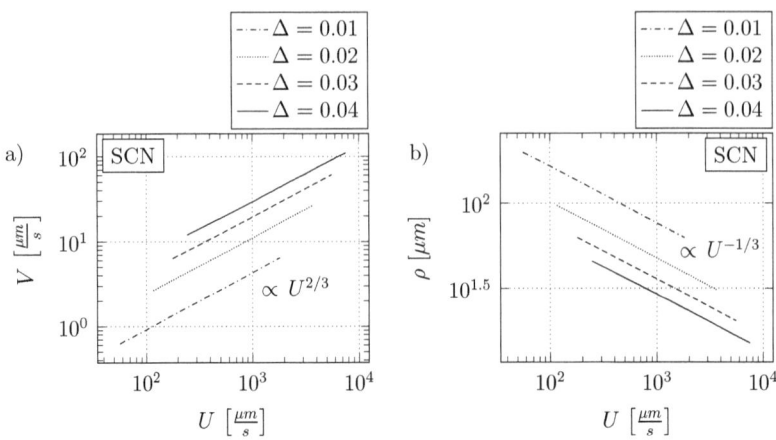

Figure 3.8.: Double logarithmic plot of a) the growth velocity V and b) the dendrite tip curvature radius ρ as functions of the forced Oseen flow velocity U determined numerically from equations (3.43) and (3.62) for succinonitrile (SCN, $\beta = 0.075$ [101]) at different values of the dimensionless undercooling Δ

this is equivalent to $\sigma = 0.0254$, which deviates somewhat from the experimental value $\sigma = 0.0195$ [64]. A reason for this deviation can be inaccuracies in the measurements of the anisotropy strength. Moreover, the eigenvalue might be corrected by including kinetic effects into the theory (see section 4.5).

Figure 3.8 shows the calculated variables V and ρ as functions of U in the range $57 \ldots 7580\,\mu m/s$, corresponding to $P_f = 0.1 \ldots 1$. The datasets are plotted for different undercoolings. The highest undercooling in use here is $\Delta = 0.04 \cong 0.92$ K, which is moderate from an experimental point of view. The double logarithmic plots reveal a power law behaviour. The approximate scaling laws

$$V \propto U^{2/3} \tag{3.79a}$$

Table 3.1.: Scaling exponents of the growth velocity V and the dendrite tip curvature radius ρ as functions of the forced Oseen flow velocity U determined from least-square fits to the data in figure 3.8 for succinonitrile at different values of the dimensionless undercooling Δ.

Δ	scaling exponent of V	scaling exponent of ρ
0.01	0.6741	-0.3340
0.02	0.6696	-0.3310
0.03	0.6650	-0.3279
0.04	0.6601	-0.3246

$$\rho \propto U^{-1/3} \tag{3.79b}$$

are found. These laws are the same as in the case of a potential flow approximation (see eq. (3.31a), (3.31b)). In the Oseen flow case, they were derived analytically only for $P_f \gg 1$ in the preceding subsection. But for the moderate flow Péclet numbers used here ($P_f = 0.1\ldots 1$ as noted above), there is no purely analytical justification of the power laws (3.79a)-(3.79b), because the Ivantsov condition (3.43) does not simplify enough in the limit $P_f \tilde{a} \gg P_c$ [123]. The results require the analytical and the numerical part of the method. The exact values of the scaling exponents determined from least-square fits to the data in figure 3.8 are given in table 3.1. The scaling laws depend only slightly on the undercooling in the probed range of $\Delta = 0.01\ldots 0.04$. However, at larger values of Δ, say $\Delta > 0.5$, they might change significantly. A tendency to this effect is visible in the table. This effect should be investigated in more detail in future work. However, at such large Δ, equation (3.43) yields $P_c > P_f \tilde{a}$ even for moderate P_f, and equation (3.62) will have to be rederived for arbitrary P_c.

If the scalings (3.79a) and (3.79b) were exact, the stability parameter would be completely independent of the forced flow velocity U. However, they are only approximate. In the probed range, σ varies between $0.025\ldots 0.0253$. This is a variation of about 1.2 % corre-

3. Convective problems

sponding to a change in U of more than one order of magnitude. For the largest value of U (at $P_f = 1.0$), σ is decreased by about 1.6 % relative to its value in the flowless case. This is in rather good agreement with the dynamical phase field simulations of Tönhardt and Amberg [120], where a variation of σ of about 2 % is observed independently of the undercooling. They simulate natural convection instead of a forced flow. But they find maximum local flow velocities of $U \approx 190 \times V$ (at $\Delta = 0.02$), which is comparable to this work. Thus, the dependence of σ on the flow is marginal, but our calculations grant access to it. This can also be seen in the local equation (3.62), where we have kept the flow term proportional to P_3. P_3 contains the asymptotically small factor σ^α, but it can be made finite by assuming $P_f \sim \sigma^{-\alpha}$.

At higher flow Péclet numbers $P_f \gtrsim 10^2$, where a comparison to the convective succinonitrile experiments of Lee et al. [84] could be made, we expect a larger change of σ as a function of U and a decrease in absolute values of the scaling exponents in (3.79a) and (3.79b). This can be seen from the following brief consideration: As noted above, σ decreases with increasing U. If this dependence becomes significant and if a power law behaviour of the form $\sigma \sim U^{-\tilde{\varepsilon}}$ with $\tilde{\varepsilon} > 0$ is assumed, then the result is $V \sim U^{(2-\tilde{\varepsilon})/3}$ and $\rho \sim U^{-(1-2\tilde{\varepsilon})/3}$. Unfortunately, we do not obtain numerical convergence in this regime. The experiments of Lee et al. were carried out with pure succinonitrile. Microscopic solvability theory based on capillary effects is known to yield more realistic results in this material than in pivalic acid. But Lee used extremely high forced flow velocities of up to $U = 1 \frac{cm}{s}$. This made a comparison impossible. The numerical procedure of this work needs to be extended to additional parameter regimes (e. g. $P_f \gg 1$). Future experiments should be carried out with pure substances such as succinonitrile and forced flow velocities in the range of thermal convection (i.e. $U \approx 10 \frac{\mu m}{s}$). A reliable value of the anisotropy strength β should be determined and a behaviour as displayed in figure 3.8 could be reproduced or disproved.

3.2. Oseen flow

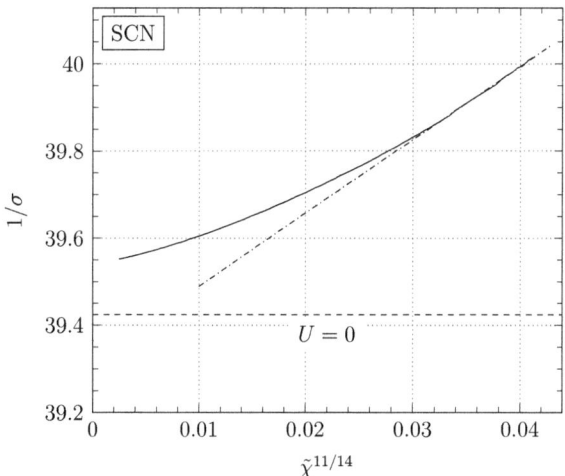

Figure 3.9.: Reciprocal stability parameter $1/\sigma$ as a function of the quantity $\tilde{\chi}^{11/14}$ with $\tilde{\chi} = \tilde{a}Ud_0/(\beta^{3/4}\rho V)$ determined numerically from equations (3.43) and (3.62) for succinonitrile ($\beta = 0.075$ [101]) in the limit $P_c \to 0$ (solid line). The dashed horizontal line indicates the value for $U = 0$. The dash-dotted line results from a least-squares fit for $\tilde{\chi}^{11/14} > 0.032$.

It should be noted, that a weak dependence of σ on P_f does not mean a weak dependence of either the growth velocity V or the tip curvature radius ρ on the externally imposed flow velocity U. The introduction of a flow changes the relationship between the undercooling and P_c via (3.43). Since P_c depends on both Δ and P_f, the same undercooling Δ will lead to a different value of P_c in the absence and in the presence of flow. As long as σ is changed only slightly, ρ will change inversely proportional to the change of P_c due to equation (1.24) and with ρ changing and $\sigma \approx$ const., V must of course also change, given that $\rho^2 V \approx$ const. according to equation (1.24).

3. Convective problems

Figure 3.9 shows the reciprocal stability parameter $1/\sigma$ as a function of the quantity $\tilde{\chi}^{11/14}$ with the dimensionless flow parameter $\tilde{\chi} = \tilde{a}Ud_0/(\beta^{3/4}\rho V)$. It is a monotonically increasing function approaching the flowless case value as $U \to 0$. The graph turns out to be independent of the undercooling. In the limit $P_f\tilde{a} \gg P_c$, Bouissou and Pelcé predict the scaling law[19]

$$\frac{1}{\sigma} \propto \tilde{\chi}^{\frac{11}{14}}. \tag{3.80}$$

The explicit dependence is given by equation (1.37). Their work was extended for arbitrary growth Péclet numbers P_c by Alexandrov and Galenko [2]. For $\tilde{\chi}^{11/14} > 0.006$, we have $P_f\tilde{a} > 10\,P_c$ and the solid graph in figure 3.9 should be a straight line if the theories are to agree. At $\tilde{\chi}^{11/14} > 0.032$, the graph becomes almost linear and theories agree well. The dash-dotted line results from a least-squares fit in this region yielding a slope of 16.80, and hence the numerical constant from equation (1.37) is $\tilde{b} = \frac{16.80}{8} \times \beta^{7/4} \approx 0.0226$ for succinonitrile. For $\tilde{\chi}^{11/14} < 0.032$, we observe a slightly curved graph. Thus, the results of the current work deviate from the scaling law (3.80) of Bouissou and Pelcé at smaller flow Péclet numbers P_f. When $\tilde{\chi}^{11/14}$ becomes significantly smaller than 0.006, equation (3.62) is less accurate.

We cannot compare our theory quantitatively with the numerical simulations of Medvedev et al. [86], because they find stability parameters of about $\sigma \approx 0.1$. In this range, $\sigma^{2/7} \approx \frac{1}{2}$ cannot be considered an asymptotically small parameter for a perturbation technique. But the analytic theory of this work is restricted to asymptotically small σ.

In order to compare the theory in this work to experiments carried out with solutions, we need to make some considerations first: The growth process is then governed by impurity diffusion instead of thermal diffusion. We follow K. Kassner [69] and write down the following

3.2. Oseen flow

chemical model:

field equations:

$$D\Delta\mu^l = \frac{\partial \mu^l}{\partial t} + (\vec{w}\cdot\vec{\nabla})\mu^l \qquad \text{in the liquid} \qquad (3.81\text{a})$$

$$D\Delta\mu^s = \frac{\partial \mu^s}{\partial t} \qquad \text{in the solid} \qquad (3.81\text{b})$$

interface conditions:

$$\mu^s = \mu^l \qquad \text{continuity} \qquad (3.82\text{a})$$

$$\mu^s = \mu_{\text{eq}} - \frac{\gamma\kappa}{\Delta c} \qquad \text{local equilibrium} \qquad (3.82\text{b})$$

$$V\vec{e}_y\cdot\vec{n} = D\left(\vec{\nabla}\mu^s - \vec{\nabla}\mu^l\right)\cdot\vec{n} \qquad \text{Stefan condition.} \qquad (3.82\text{c})$$

Here, the chemical potential μ is the relevant field quantity. μ_{eq} is the equilibrium value at the phase transition and Δc is the miscibility gap at concentration c. D is the impurity diffusivity here. The flow velocity field can be calculated from the Oseen equation (3.35) with interface conditions (1.13a)-(1.13b). The dimensionless chemical potential field

$$u = \frac{1}{P_c}\frac{\mu - \mu_{\text{eq}}}{\Delta c \frac{\partial \mu}{\partial c}} \qquad (3.83)$$

is introduced. The capillary length is set $d_0 = \gamma\left/\left((\Delta c)^2 \frac{\partial \mu}{\partial c}\right)\right.$. The time, the flow velocity as well as all lengths are scaled as given in section 2.1. Moreover, a moving frame of reference is used, moving along with the dendrite, which is assumed to grow at constant velocity V:

field equations:

$$\Delta u^l - (\vec{w}\cdot\vec{\nabla})u^l = 0 \qquad \text{in the liquid} \qquad (3.84\text{a})$$

$$\Delta u^s + P_c(\vec{e}_y\cdot\vec{\nabla})u^s = 0 \qquad \text{in the solid} \qquad (3.84\text{b})$$

3. Convective problems

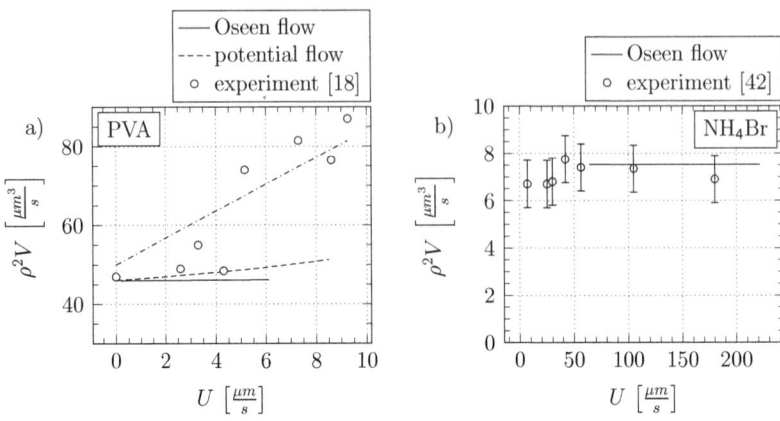

Figure 3.10.: Oseen flow: Comparison to experiments with two-component systems; denominator $\rho^2 V$ of the stability parameter as a function of the forced flow velocity U, a) 2 % ethanol in pivalic acid at $\Delta = 0.0169$, $\beta = 0.09$, $d_0 = 32\,\text{Å}$ to be compared to [18] (circles and dash-dotted line), b) solid line: 51 % NH$_4$Br solution in water at $\Delta = 0.01$, circles: measured data for 43 % NH$_4$Br solution in water from the experiments of Emsellem and Tabeling [42]

interface conditions:

$$u^s = u^l \qquad \text{continuity} \qquad (3.85\text{a})$$

$$u^s = -\frac{1}{2}\sigma\kappa a(\theta) \qquad \text{local equilibrium} \qquad (3.85\text{b})$$

$$\vec{e}_y \cdot \vec{n} = \left(\vec{\nabla}u^s - \vec{\nabla}u^l\right) \cdot \vec{n} \qquad \text{Stefan condition.} \qquad (3.85\text{c})$$

By renaming $u \to T$, this model becomes mathematically equivalent to the thermal model used in this chapter. The calculations and equations from the preceding sections can be used. However, an issue about the chemical model has to be noted: The impurity diffusivity is known to be orders of magnitude smaller in the solid phase than in the liquid phase [82]. Assuming D to be equal in both domains is a much rougher approximation in the chemical model than in the thermal model. Strictly speaking, a one-sided model would have to be used, where the heat balance term corresponding to u^s in the Stefan condition (3.85c) is dropped. In summary, the following statements concerning the comparison between this work and experiments with two-component systems should rather be considered as an estimate. An alternative approach for convective systems is exhibited in a recent theoretical article by Alexandrov et al. [3], which constitutes an extension of the linear-stability-analysis-based work of Bouissou and Pelcé [19] to binary systems.

There is no good agreement between this work and the experiments by Bouissou et al. carried out with pivalic acid [18] (see fig. 3.10a). They use 2 % - 4 % ethanol in their pivalic-acid-based working substance. Here things are compared to their 2 %-results. Thus, D was set to 250 $\frac{\mu m^2}{s}$ and d_0 was now set to 32 Å for the calculation of the data (the pure substance value is 37.6 Å [101]). It can be seen, that we are in the right range. However, one cannot say, that there is a good agreement between theory and experiment at this point. Our data varies much less with U than the experiments imply. $\rho^2 V$ exhibits a relatively marginal dependence on the forced flow velocity despite $P_f \tilde{a} \gg P_c$. Thus, an imposed flow far ahead of the interface has almost no impact on the eigenvalue in the considered parameter range. In this respect, the potential flow yields better results for the pivalic acid solution, although it is supposed to be the less accurate approximation. The discrepancy might occur partially because of kinetic interface effects, which were neglected. This is discussed in more detail for the flowless case in section 4.5. Moreover, the real flow velocity field may deviate considerably from the approximations used here. In addition to that, as mentioned above, the experiments in [18] do not exactly realize the situation described in this work. Impurity diffusion is much slower in the solid than it was assumed here, and the solution in the solid probably deviates from the real situation. This is

3. Convective problems

another reason for the discrepancies.

In contrast to that, there is decent agreement between this work and the experiments of Emsellem et al. carried out using a 43 % ammonium bromide (NH$_4$Br) solution in water at an undercooling of $\Delta \approx 0.01$ [42] (fig. 3.10b). The numerical data is calculated for 51 % NH$_4$Br solution in water by weight (because it seems like Emsellem et al. simply have not measured D and d_0 for their concentration). Additional experimental work on dendritic growth in this substance can be found in [63]. The viscosity of the solution does not differ much from that of pure water at 25 °C. The important material parameters of ammonium bromide are shown below. For an overview over the material parameters of all the substances considered in this work, see appendix C.

ammonium bromide (NH$_4$Br) solution in water:

- $\beta = 0.24$ [40]
- $D = 2600\,\mu m^2/s$ [40]   } for 51 % NH$_4$Br solution in water by weight
- $d_0 = 0.28\,nm$ [40]
- $\nu = 8.931 \cdot 10^5\,\mu m^2/s$ [42]   taken from pure water at 25 °C
- $Pr = 343.5$ ν/D from the values above

Despite the finite difference between the concentrations used in the calculation and the experiment, there is rather good agreement in the plotted range. The fact that microscopic solvability theory yields more realistic results for NH$_4$Br and works worse for pivalic acid and succinonitrile has already been mentioned before [91]. We find, that $\rho^2 V$ is almost constant at 7.52 $\frac{\mu m^3}{s}$. Accordingly, the forced flow has nearly no effect on the selected stability parameter σ because $P_3 \ll 1$ in our data.

Finally, the question about the applicability of the Oseen approximation remains open because, as mentioned above, for experimentally relevant substances, pivalic acid as well as succinonitrile, at $\Delta \ll 1$ we are in the limit $P_f \tilde{a} \gg P_c$. This is the very limiting case, in which the zeroth order approximation to the flow velocity field is expected to become inaccurate. Despite $P_f \tilde{a} \gg P_c$, we do not see a large effect of the forced flow. But if we had taken into account the terms of P_c/P_r at each stage of the calculations, we would have ended up in the case of arbitrary growth Péclet numbers in the Zauderer scheme, which is not what we wanted to consider here. This is discussed separately for the flowless case in section 4.4.

4. Extended heat transport properties

4.1. Nonlinear diffusion

4.1.1. Ivantsov solution for dendritic growth in a system with temperature-dependent thermal diffusivity

Dendritic growth with a temperature-dependent thermal diffusivity $D(T)$ was considered by Kurtze [76]. He reduced the free boundary problem to the solution of one single one-dimensional ordinary differential equation. $D(T)$ is now a function of the dimensionless temperature. It is factorized into a constant dimensional part and a dimensionless function f_D of T.

$$D(T) = D \cdot f_D(T) \tag{4.1}$$

Neglecting convection, the diffusion equation (1.8a) is used to describe heat transport in the liquid and in the solid domain. I.e. we treat both phases symmetrically. Regarding the factorization (4.1) and applying the product rule on the right hand side of (1.8a), one finds the dimensionless equation

$$f'_D \left|\vec{\nabla} T\right|^2 + f_D \Delta T + P_c \left(\vec{e}_y \cdot \vec{\nabla}\right) T = 0 \tag{4.2}$$

4. Extended heat transport properties

with the P_c-term resulting from the moving frame of reference attached to the interface. Again, we look for stationary solutions in this frame. The prime denotes the derivative with respect to the dimensionless temperature T. Equation (4.2) is nonlinear and therefore, its solution might constitute an unsurmountable obstacle. The boundary conditions (1.9a)-(1.9c) are rewritten using the non-dimensionalizations from section 2.1, neglecting the kinetic term. Subsequently, the equations, including (4.2), are transformed to conformal parabolic coordinates and we find a convenient form of the problem to be investigated in this section:

diffusion equation:
$$0 = f'_D \left(T_\xi^2 + T_\eta^2\right) + f_D \left(T_{\xi\xi} + T_{\eta\eta}\right) \\ + P_c \left(\eta T_\eta - \xi T_\xi\right) \quad \text{in both phases} \quad (4.3)$$

interface conditions:
$$T^l = T^s \qquad \text{continuity} \quad (4.4a)$$
$$T^s = -\frac{1}{2}\sigma \kappa a(\theta) \qquad \text{Gibbs-Thomson} \quad (4.4b)$$
$$[\xi \eta_s]' = f_D(T) \left(\partial_\eta - \eta'_s \partial_\xi\right) \left(T^s - T^l\right) \qquad \text{Stefan condition.} \quad (4.4c)$$

Looking for a similarity solution to the problem (4.3)-(4.4c) in the case of vanishing surface tension ($\sigma = 0$), we require $T = T(\eta)$. Equation (4.3) becomes

$$0 = \frac{\mathrm{d}}{\mathrm{d}\eta}\left(f_D T_\eta\right) + P_c \eta T_\eta \qquad (4.5)$$

in the liquid. The boundary conditions (4.4b) and (4.4c) read $T(1) = 0$ and $T_\eta(1) = -1/f_D(0)$ respectively. Setting the solid isothermal ($T^{s,\mathrm{lv}} = 0$), the continuity condition (4.4a) is fulfilled automatically. After dividing both sides by f_D, (4.5) can be solved by separation of variables,

$$T^{\text{Iv}}(\eta) = -\int_1^\eta \frac{1}{f_D(T^{\text{Iv}}(\eta'))} \exp\left(-P_c \int_1^{\eta'} \frac{\eta''}{f_D(T^{\text{Iv}}(\eta''))} d\eta''\right) d\eta'. \quad (4.6)$$

This already fulfills the boundary conditions. It is an implicit form of the solution. For a given algebraic form of $f_D(T)$, equation (4.5) might not be integrable in every case due to its nonlinearity. The Ivantsov condition is

$$\Delta = \int_1^\infty \frac{P_c}{f_D(T^{\text{Iv}}(\eta'))} \exp\left(-P_c \int_1^{\eta'} \frac{\eta''}{f_D(T^{\text{Iv}}(\eta''))} d\eta''\right) d\eta'. \quad (4.7)$$

In order to obtain a temperature field correction at finite σ, we make the transition $T \to T + T^{\text{Iv}}$ and the function $f_D(T + T^{\text{Iv}})$ is expanded in a Taylor series about T^{Iv}:

$$f_D(T + T^{\text{Iv}}) = f_D(T^{\text{Iv}}) + f_D'(T^{\text{Iv}})T. \quad (4.8)$$

From here on, we will use the short forms

$$f_D(T^{\text{Iv}}) = f_0 \qquad\qquad f_D'(T^{\text{Iv}}) = f_0' \quad (4.9)$$

for $f_D(T^{\text{Iv}})$ and $f_D'(T^{\text{Iv}})$. The expanded diffusion equation is derived starting from the non-dimensional form of (4.2) in the moving frame of reference:

$$\begin{aligned}
0 =\ & P_c(\vec{e}_y \cdot \vec{\nabla})(T + T^{\text{Iv}}) + \vec{\nabla}\left((f_0 + f_0'T)\vec{\nabla}(T + T^{\text{Iv}})\right) \\
=\ & P_c(\vec{e}_y \cdot \vec{\nabla})(T + T^{\text{Iv}}) + (f_0 + f_0'T)\Delta(T + T^{\text{Iv}}) \\
& + \left(f_0'\vec{\nabla}T^{\text{Iv}} + f_0''T\,\vec{\nabla}T^{\text{Iv}} + f_0'\vec{\nabla}T\right)\vec{\nabla}(T + T^{\text{Iv}}) \\
=\ & P_c(\vec{e}_y \cdot \vec{\nabla})(T + T^{\text{Iv}}) + (f_0 + f_0'T)\Delta(T + T^{\text{Iv}}) \\
& + f_0'|\vec{\nabla}(T + T^{\text{Iv}})|^2 + Tf_0''\left(|\vec{\nabla}T^{\text{Iv}}|^2 + \vec{\nabla}T \cdot \vec{\nabla}T^{\text{Iv}}\right)
\end{aligned}$$

4. Extended heat transport properties

$$\begin{aligned}
\Rightarrow 0 =\ & P_c\left(\eta(T_\eta + T_\eta^{\text{Iv}}) - \xi T_\xi\right) + (f_0 + f_0'T)\left(T_{\xi\xi} + T_{\eta\eta} + T_{\eta\eta}^{\text{Iv}}\right) \\
& + f_0'\left(T_\xi^2 + T_\eta^2 + 2T_\eta T_\eta^{\text{Iv}} + T_\eta^{\text{Iv}\,2}\right) + T f_0''\left(T_\eta^{\text{Iv}\,2} + T_\eta T_\eta^{\text{Iv}}\right) \\
=\ & P_c\left(\eta T_\eta - \xi T_\xi\right) + (f_0 + f_0'T)\left(T_{\xi\xi} + T_{\eta\eta}\right) \\
& + f_0'\left(T_\xi^2 + T_\eta^2 + 2T_\eta T_\eta^{\text{Iv}} + T T_{\eta\eta}^{\text{Iv}}\right) + T f_0''\left(T_\eta^{\text{Iv}\,2} + T_\eta T_\eta^{\text{Iv}}\right) \\
& \qquad\qquad\qquad \underbrace{\phantom{P_c\eta T_\eta^{\text{Iv}} + f_0'T_\eta^{\text{Iv}\,2} + f_0 T_{\eta\eta}^{\text{Iv}}}}_{=0 \text{ (see (4.5))}} \\
& + \cancel{P_c\eta T_\eta^{\text{Iv}}} + \cancel{f_0'T_\eta^{\text{Iv}\,2}} + f_0 T_{\eta\eta}^{\text{Iv}}.
\end{aligned}$$

Hence, the expanded diffusion equation is

$$\begin{aligned}
0 =\ & P_c\left(\eta T_\eta - \xi T_\xi\right) + (f_0 + f_0'T)\left(T_{\xi\xi} + T_{\eta\eta}\right) \\
& + f_0'\left(T_\xi^2 + T_\eta^2 + 2T_\eta T_\eta^{\text{Iv}} + T T_{\eta\eta}^{\text{Iv}}\right) + T f_0''\left(T_\eta^{\text{Iv}\,2} + T_\eta T_\eta^{\text{Iv}}\right)
\end{aligned} \qquad (4.10)$$

in both phases. The interface is now located at $\eta_s = 1+h(\xi)$. $T^{\text{Iv}}(1+h)$ and $T_\eta^{\text{Iv}}(1+h)$ in the liquid are expanded in a Taylor series about $\eta = 1$. The boundary conditions (3.10b) and (3.10d) can be used. Because of $T^{\text{Iv}}(1) = 0$ and $T_\eta^{\text{Iv}}(1) = -1/f_D(0)$, the expanded continuity condition is

$$T^l = T^s + \frac{h}{f_D(0)}. \qquad (4.11)$$

Now the Stefan condition (4.4c) is expanded:

$$\xi h' + 1 + h = \left(f_D(0) + f_D'(0)T\right) \\
\times \left[(\partial_\eta - h'\partial_\xi)\left(T^s - T^l\right) - T_\eta^{\text{Iv}}(1) - T_{\eta\eta}^{\text{Iv}}(1)h\right].$$

From (4.5), we find

$$\begin{aligned}
T_{\eta\eta}^{\text{Iv}}(1) &= -\frac{1}{f_D(0)}\left(f_D'(0)T_\eta^{\text{Iv}\,2}(1) + P_c T_\eta^{\text{Iv}}(1)\right) \\
&= -\frac{1}{f_D(0)}\left(\frac{f_D'(0)}{f_D^2(0)} - \frac{P_c}{f_D(0)}\right).
\end{aligned}$$

This is inserted into the expanded Stefan condition and after eliminating the 1 on the left hand side with the $-f_D(0)T_\eta^{\text{Iv}}(1)$-term from the right hand side, we arrive at

$$[\xi h]' = \left(f_D(0) + f'_D(0)T\right)(\partial_\eta - h'\partial_\xi)\left(T^s - T^l\right)$$
$$+ \frac{f'_D(0)}{f_D(0)}T + \frac{h}{f_D^2(0)}\left(f'_D(0) - P_c f_D(0)\right) \quad (4.12)$$

for the expanded Stefan condition. Terms of the order $\mathcal{O}(Th)$ were neglected here.

4.1.2. Continuation to the complex plane and asymptotic decomposition (nonlinear Diffusion)

We would like to decompose equation (4.10) with conditions (4.11)-(4.12), (3.10b) and (3.10d) à la Zauderer. In contrast to section 2.2, we employ the three-component variable and the corresponding 3×3-matrices

$$\vec{\vartheta} = (T_\xi, T_\eta, T)^\mathrm{T} \quad (4.13)$$

$$E = \begin{pmatrix} 0 & 1 & 0 \\ -1 & 0 & 0 \\ 0 & 0 & 1 \end{pmatrix} \qquad F = \begin{pmatrix} -u_2 & -v_2 & -w_2 \\ 0 & 0 & 0 \\ -1 & -1 & 0 \end{pmatrix} \quad (4.14)$$

instead of two-dimensional quanities to write (4.10) as a first order system.

$$\vec{\vartheta}_\xi + E\vec{\vartheta}_\eta + F\vec{\vartheta} = 0 \quad (4.15)$$

The functions u_2, v_2 and w_2 in F are obtained by rearranging the nonlinear equation (4.10). w_2 must contain all expressions multiplying T. Then, T_ξ and T_η must be factored out in the remaining terms for u_2 and v_2 respectively:

$$u_2 = -\frac{1}{f_0 + f_0''T}\left(f_0''T_\xi - P_c\xi\right) \quad (4.16a)$$

4. Extended heat transport properties

$$v_2 = -\frac{1}{f_0 + f_0''T} \left(f_0' T_\eta + P_c \eta + 2 f_0' T_\eta^{\text{Iv}} \right) \tag{4.16b}$$

$$w_2 = -\frac{1}{f_0 + f_0''T} \left(f_0' T_{\eta\eta}^{\text{Iv}} + f_0'' T_\eta^{\text{Iv}\,2} + f_0''' T_\eta T_\eta^{\text{Iv}} \right). \tag{4.16c}$$

The eigenvectors $\vec{r}_{3,4,5}$ of E, corresponding to the eigenvalues i, $-$i and 1 respectively,

$$\vec{r}_3 = \begin{pmatrix} -\text{i} \\ 1 \\ 0 \end{pmatrix} \qquad \vec{r}_4 = \begin{pmatrix} \text{i} \\ 1 \\ 0 \end{pmatrix} \qquad \vec{r}_5 = \begin{pmatrix} 0 \\ 0 \\ 1 \end{pmatrix} \tag{4.17}$$

are used to expand $\vec{\vartheta}$ in the liquid and in the solid:

$$\vec{\vartheta} = M\vec{r}_3 + \varepsilon N\vec{r}_4 + Q\vec{r}_5, \tag{4.18a}$$

$$\vec{\vartheta}^s = \varepsilon M^s \vec{r}_3 + N^s \vec{r}_4 + Q^s \vec{r}_5. \tag{4.18b}$$

These linear combinations contain the identity $Q = T$. Then, from the Gibbs-Thomson condition (3.10b) it is obvious, that Q must diverge at the singularity at $\xi = -\text{i}$. For this reason, it must not be multiplied by a factor ε in (4.18a) and (4.18b). However, the derivatives T_ξ and T_η diverge faster than Q at the singularity. (4.18a), (4.18b) are inserted into (4.15), and after the scale transformation $\xi, \eta \to \varepsilon\xi, \varepsilon\eta$ the equation becomes

$$(M_\xi + \text{i}M_\eta)\vec{r}_3 + \varepsilon(N_\xi - \text{i}N_\eta)\vec{r}_4 + (Q_\xi + Q_\eta)\vec{r}_5 \\ + \varepsilon F M \vec{r}_3 + \varepsilon F Q \vec{r}_5 = 0 \tag{4.19a}$$

$$\varepsilon(M_\xi^s + \text{i}M_\eta^s)\vec{r}_3 + (N_\xi^s - \text{i}N_\eta^s)\vec{r}_4 + (Q_\xi^s + Q_\eta^s)\vec{r}_5 \\ + \varepsilon F N^s \vec{r}_4 + \varepsilon F Q^s \vec{r}_5 = 0 \tag{4.19b}$$

in the liquid and in the solid respectively, neglecting terms of the order $\mathcal{O}(\varepsilon^2)$. Equation (4.19a) is projected onto the invariant subspaces of

4.1. Nonlinear diffusion

E. The corresponding projection operators are

$$P_3 = \frac{1}{2}\begin{pmatrix} 1 & -i & 0 \\ i & 1 & 0 \\ 0 & 0 & 0 \end{pmatrix}, \quad P_4 = \frac{1}{2}\begin{pmatrix} 1 & i & 0 \\ -i & 1 & 0 \\ 0 & 0 & 0 \end{pmatrix}, \quad P_5 = \begin{pmatrix} 0 & 0 & 0 \\ 0 & 0 & 0 \\ 0 & 0 & 1 \end{pmatrix}.$$
(4.20)

We use use the formulas

$$P_3 F \vec{r}_3 = -\frac{1}{2}(u_2 + iv_2)\vec{r}_3 \qquad P_3 F \vec{r}_5 = -\frac{i}{2}w_2 \vec{r}_3 \qquad (4.21a)$$

$$P_4 F \vec{r}_3 = \frac{1}{2}(u_2 + iv_2)\vec{r}_4 \qquad P_4 F \vec{r}_5 = \frac{i}{2}w_2 \vec{r}_4 \qquad (4.21b)$$

$$P_5 F \vec{r}_3 = (i-1)\vec{r}_5 \qquad P_5 F \vec{r}_5 = 0 \qquad (4.21c)$$

and find after projection and returning to the original scale

$$M_\xi + iM_\eta = \frac{1}{2}(u_2 + iv_2)M + \frac{i}{2}w_2 Q \qquad (4.22a)$$

$$N_\xi - iN_\eta = -\frac{1}{2}(u_2 + iv_2)M - \frac{i}{2}w_2 Q \qquad (4.22b)$$

$$Q_\xi + Q_\eta = (1-i)M \qquad (4.22c)$$

in the liquid. Equations (4.16a)-(4.16b) are used to write the function $u_2 + iv_2$ explicitly:

$$u_2 + iv_2 = -\frac{1}{f_0 + f_0'T}\left[-P_c(\xi - i\eta) + 2if_0''T_\eta^{\text{Iv}} + f_0'(T_\xi + iT_\eta)\right].$$

The linear combination (4.18a) leads to $T_\xi = -i(M - N)$ and $T_\eta = M + N$. Thus, $T_\xi + iT_\eta = 2iN$ can be inserted into $u_2 + iv_2$ and $T_\eta = M + N$ is used in w_2 from (4.16c). Furthermore, in w_2 the identity $f_0'T_{\eta\eta}^{\text{Iv}} + f_0''T_\eta^{\text{Iv2}} = \frac{d}{d\eta}\left(f_0'T_\eta^{\text{Iv}}\right)$ is used. Equations (4.22a)-(4.22b)

4. Extended heat transport properties

become

$$M_\xi + iM_\eta = \frac{1}{f_0 + f_0'Q} \left[\left(\frac{P_c}{2}(\xi - i\eta) - if_0'\left(N + T_\eta^{\text{Iv}}\right) \right) M \right. \\ \left. - \frac{i}{2} \left(f_0'' T_\eta^{\text{Iv}}(M+N) + \frac{d}{d\eta}\left(f_0' T_\eta^{\text{Iv}}\right) \right) Q \right], \quad (4.23a)$$

$$N_\xi - iN_\eta = -\frac{1}{f_0 + f_0'Q} \left[\left(\frac{P_c}{2}(\xi - i\eta) - if_0'\left(N + T_\eta^{\text{Iv}}\right) \right) M \right. \\ \left. - \frac{i}{2} \left(f_0'' T_\eta^{\text{Iv}}(M+N) + \frac{d}{d\eta}\left(f_0' T_\eta^{\text{Iv}}\right) \right) Q \right]. \quad (4.23b)$$

The coupled system of differential equations (4.23a)-(4.23b), (4.22c) is nonlinear and searching for an analytical solution is quite futile. However, things can be simplified to a certain amount. First, we take the limit $P_c \to 0$ as in chapter 3, where dendritic growth with convection is treated. Second, we neglect nonlinear terms. This leaves us with equations, which are not correct up to the first order in $\sigma^{2/7}$ anymore. But the effect of the nonlinear diffusion is still present, because the function f_0' is kept, multiplying the derivative T_η^{Iv} of the solution of the unperturbed problem,

$$M_\xi + iM_\eta = -\frac{i}{f_0} \left[f_0' T_\eta^{\text{Iv}} M + \frac{1}{2} \frac{d}{d\eta}\left(f_0' T_\eta^{\text{Iv}}\right) Q \right], \quad (4.24a)$$

$$N_\xi - iN_\eta = \frac{i}{f_0} \left[f_0' T_\eta^{\text{Iv}} M + \frac{1}{2} \frac{d}{d\eta}\left(f_0' T_\eta^{\text{Iv}}\right) Q \right]. \quad (4.24b)$$

For constant D, we recover (2.23a) and (2.23b). The system (4.24a)-(4.24b), (4.22c) determines the approximate temperature field correction in the liquid. The corresponding equations in the solid can be derived analogously by applying $P_{3,4,5}$ from the left to equation (4.19b). But it is not clear, how the system (4.24a)-(4.24b), (4.22c) can be solved. Despite the linearization and the limit $P_c \to 0$, the equations are still coupled. This problem will not disappear, if the temperature dependence of the thermal diffusivity is given explicitly. For instance, assuming the linear limit $f_D(T) \propto T$, the Ivantsov integral remains analytically unsolvable and the issue persists. And even if a solution could be found, the boundary conditions (4.11)-(4.12), (3.10b) and (3.10d) would have to be decomposed and applied subsequently, which would constitute another mathematical obstacle, since there are six coefficient functions to be dealt with here. In summary, the decomposed equations for the temperature field correction were derived here, but a dendritic growth mode in systems with nonlinear diffusion cannot be determined in this work. To progress further, an expansion for small deviations from $D(T) = $ const. could be tried.

4.2. Thermal resistance at the two-phase boundary

A finite thermal resistance at the interface of the crystal shall be investigated in this section. Its effect on dendritic growth is negligible in common substances and at common working temperatures. However, its influence may be significant in systems such as liquid ^{3}He [56] or alloys at superfluid helium temperatures [94]. Imagine a copper block immersed in liquid helium. The block is heated with constant power P, and it takes an equilibrium temperature T_1, after the heat fluxes have balanced. Now, a second copper block is dipped into the fluid without external heating. Heat from the first block flows through the fluid to the second block, which takes a steady-state temperature T_2. The system is illustrated in figure 4.1. Since T_2 remains smaller than T_1 even after long times, there must be some thermal resistance $R_K = (T_1 - T_2)/P$ in the system. Surprisingly enough, the equilibrium temperature T_2 turns out to be independent of the distance d between the two copper blocks. Thus, the thermal resistance cannot result from the fluid, but it must be an effect of the phase boundary.

4. Extended heat transport properties

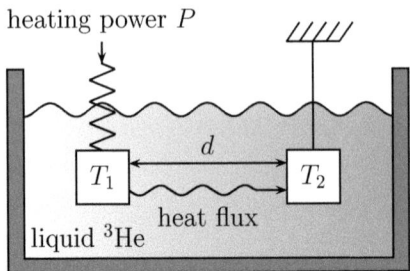

Figure 4.1.: Two copper blocks immersed in a liquid ^{3}He bath. The heating of block 1 with constant power P causes a heat flux to block 2 at distance d. In equilibrium, one has $T_1 > T_2$ due to the thermal resistance at the copper-^{3}He interfaces.

The phenomenon is also referred to as *Kapitza effect*. It influences the dendritic growth mode, and the model equations from section 1.1 must be altered. The phases cannot be treated symmetrically anymore, because the thermal diffusivity D_s in the solid is significantly larger than the thermal diffusivity D_l in liquid ^{3}He[1]. Their ratio is $\mu_K = D_s/D_l \approx 220$ at $T = 0.1$ K [99]. This fact is unusual, but at the interface an effective ratio can be introduced, which is less than unity, because according to Graner, Bowley and Nozières [57] and Balibar, Edwards and Saam [12], all latent heat is released into the liquid, and this latent heat has to overcome the thermal resistance before it can diffuse into the solid [99]. Allowing for different thermal diffusivities complicates the problem mathematically. For that reason, convection is neglected in this section, since we are interested in solving a model, which is as simple as possible (but as accurate as necessary). Yet, flow effects should be more or less easy to consider, if the helium is superfluid, because a mere potential flow is a good approximation in this system due to the vanishing viscosity.

---

[1] Note, that it is the other way around in *superfluid* ^{3}He, where D_l is much larger than D_s.

4.2.1. Ivantsov solution for dendritic growth with thermal interface resistance

In a frame of reference moving at constant growth velocity $V\vec{e}_y$, the dimensionless and convection-free forms of the bulk equations (1.8a) and (1.8b) read

$$\Delta T^l + P_c(\vec{e}_y \cdot \vec{\nabla})T^l = 0 \qquad \text{in the liquid} \qquad (4.25a)$$

$$\mu_K \Delta T^s + P_c(\vec{e}_y \cdot \vec{\nabla})T^s = 0 \qquad \text{in the solid} \qquad (4.25b)$$

with $P_c = \rho V/D_l$. Again, the non-dimensionalizations can be found in section 2.1. Due to the thermal resistance, the temperature is not continuous at the two-phase boundary anymore. The (dimensional) temperature difference at the interface is

$$T^l - T^s = R_K(J_E - \tilde{\lambda}J) \qquad (4.26)$$

with the total heat current $J_E = Q_s + TS_sJ = Q_l + TS_lJ$ through the interface and the mass current J through the interface [30]. Here, $Q_{s,l}$ are the conductive heat flows in the solid and the liquid phases respectively, and $S_{s,l}$ are the entropies of each phase [99]. J_E is inserted into the temperature difference:

$$T^l - T^s = R_K \left[Q_s + (TS_s - \tilde{\lambda})J \right]. \qquad (4.27)$$

The coefficient $\tilde{\lambda}$ determines, which fraction of each entropy is liberated or absorbed on each side of the interface. Since all latent heat is released into the liquid, the cross coefficient is $\tilde{\lambda} = TS_s$, and only $R_K Q_s$ remains on the right hand side of (4.27) [99]. The conductive heat flow in the solid is $Q_s = K_s \vec{\nabla} T^s \cdot \vec{n}$ with the solid heat conductivity $K_s = \varrho_m c D_s$. The interface unit normal vector $\vec{n}$ points into the liquid again. The dimensionless boundary conditions read:

$$T^l - T^s = \gamma_K \vec{\nabla} T^s \cdot \vec{n} \qquad \text{therm. discontinuity} \qquad (4.28a)$$

4. Extended heat transport properties

$$T^l = -\frac{1}{2}\sigma\kappa a(\theta) \qquad \text{Gibbs-Thomson} \qquad (4.28b)$$

$$\vec{e}_y \cdot \vec{n} = \left(\mu_K \vec{\nabla} T^s - \vec{\nabla} T^l\right) \cdot \vec{n} \quad \text{asymmetric Stefan condition.} \quad (4.28c)$$

Besides, far ahead of the interface, a fixed undercooling is required again: $T^l \to -\Delta/P_c$. The parameter group γ_K is defined by

$$\gamma_K = \frac{R_K K_s}{\rho} = \frac{R_K K_s \sigma P_c}{2 d_0} \qquad (4.29)$$

where R_K is the Kapitza resistance measured in units of $K \cdot m^2/W$. Rolley et al. estimated $R_K K_s \approx 3\, mm$ for liquid ^{3}He at $T = 0.1$ K [99]. There is no $\vec{\nabla} T^l$-term on the right hand side of (4.28a) because of the cross coefficient $\tilde{\lambda} = T S_s$. Kinetic effects were neglected in the Gibbs-Thomson condition (4.28b). The important material parameters of ^{3}He are shown below.

liquid ^{3}He at T=0.1 K:

- $\beta = 0.3$ [99]
- $d_0 = 3.8\, nm$ [99]
- $\gamma = 6 \cdot 10^{-5}\, \frac{J}{m^2}$ [98]
- $D_l = 5.46 \cdot 10^4\, \frac{\mu m^2}{s}$ [59]
- $c = 719\, \frac{J}{kg \cdot K}$ [58]
- $R_K = 1\, \frac{cm^2 \cdot K}{W}$ [56]
- $R_K K_s = 3000\, \mu m$ [99]
- $\mu_K = 220$ [99], [59]
- $T_M = 0.32\, K$ [99]

4.2. Thermal resistance at the two-phase boundary

The equation for the shape correction function is calculated in cartesian *and* parabolic coordinates in this section. Therefore, we started with cartesian model equations. In parabolic coordinates, the model (4.25a)-(4.25b), (4.28a)-(4.28c) reads

field equations:
$$T^l_{\xi\xi} + T^l_{\eta\eta} + P_c(\eta T^l_\eta - \xi T^l_\xi) = 0 \quad \text{in the liquid} \quad (4.30a)$$
$$\mu_K(T^s_{\xi\xi} + T^s_{\eta\eta}) + P_c(\eta T^s_\eta - \xi T^s_\xi) = 0 \quad \text{in the solid} \quad (4.30b)$$

interface conditions:
$$(T^l - T^s)\sqrt{(\xi^2 + \eta_s^2)(1+\eta_s'^2)} \quad \text{therm. dis-} \\ = \gamma_K\left(T^s_\eta - \eta_s' T^s_\xi\right) \quad \text{continuity} \quad (4.31a)$$
$$T^l = -\frac{1}{2}\sigma\kappa a(\theta) \quad \text{Gibbs-Thomson} \quad (4.31b)$$
$$[\xi\eta_s]' = (\partial_\eta - \eta_s'\partial_\xi)\left(\mu_K T^s - T^l\right) \quad \text{asym. Stefan.} \quad (4.31c)$$

In the experiments of Rolley et al. [99], the time scale of heat transport in the solid was much shorter than that of the growth process itself. I.e. the latent heat in the solid was removed almost instantly causing any thermal gradients in this domain to be very small. Thus, assuming the solid to be isothermal is not a bad approximation. For $\sigma = 0$, the temperature depends only on η. Using this information, the form

$$T^{l,\mathrm{Iv}}(\eta) = -\int_1^\eta e^{-\frac{P_c}{2}(\omega^2-1)}d\omega \quad (4.32)$$

is a solution to equation (4.30a), and it already fulfills the interface conditions (4.31a)-(4.31c). Similar calculations were carried out in subsection 3.1.1. The temperature in the whole solid domain is just $T^{s,\mathrm{Iv}} = 0$. Taking into account the far field boundary condition, the Ivantsov condition (1.17) holds. This would be more complicated, if a

4. Extended heat transport properties

$\vec{\nabla}T^l$-term was taken into account on the right hand side of the thermal discontinuity condition (4.28a).

The model equations are expanded about the isothermal interface solution (4.32). The bulk equations are linear and apply to all orders of the perturbation solution. We take the limit $P_c \to 0$,

$$T^{l,s}_{\xi\xi} + T^{l,s}_{\eta\eta} = 0. \tag{4.33}$$

Again, setting $\eta_s = 1 + h(\xi)$ and $T \to T + T^{\mathrm{Iv}}$, the solution (4.32) and its derivative are expanded about $\eta = 1$:

$$T^{l,\mathrm{Iv}}(1+h) \approx \underbrace{T^{l,\mathrm{Iv}}(1)}_{=0} + \underbrace{T^{l,\mathrm{Iv}}_\eta(1)}_{=-1} h = -h, \tag{4.34a}$$

$$T^{l,\mathrm{Iv}}_\eta(1+h) \approx \underbrace{T^{l,\mathrm{Iv}}_\eta(1)}_{=-1} + \underbrace{\frac{\mathrm{d}T^{l,\mathrm{Iv}}_\eta}{\mathrm{d}\eta}\bigg|_1}_{\propto P_c} h \approx -1. \tag{4.34b}$$

Inserting into (4.31a)-(4.31c) yields

$$(T^l - T^s - h)\sqrt{(\xi^2 + (1+h)^2)(1+h'^2)} \quad \text{therm. dis-}$$
$$= \gamma_K \left(T^s_\eta - h' T^s_\xi\right) \quad \text{continuity} \tag{4.35a}$$

$$T^l = -\frac{1}{2}\sigma \kappa a(\theta) + h \quad \text{Gibbs-Thomson} \tag{4.35b}$$

$$[\xi h]' = (\partial_\eta - h'\partial_\xi)\left(\mu_K T^s - T^l\right) \quad \text{asym. Stefan.} \tag{4.35c}$$

The system (4.33), (4.35a)-(4.35c) determines the correction to the temperature field.

4.2.2. Shape equation in parabolic and cartesian coordinates for dendritic growth with finite thermal resistance

As discussed in section 2.2, in the case of purely diffusive heat transport and vanishing growth Péclet number P_c, we may use

$$T^l_\xi = -iT^l_\eta \tag{4.36a}$$
$$T^s_\xi = iT^s_\eta \tag{4.36b}$$

close to $\xi = -i$, because (4.33) is a Laplace equation. We do not need to calculate the solutions explicitly. Equations (4.36a) and (4.36b) provide enough information to eliminate $T^{l,s}$ and its derivatives from the system of interface equations (4.35a)-(4.35c). $T^{l,s}_\eta$ in (4.35c) is replaced using (4.36a) and (4.36b). One finds

$$i\left[\xi h\right]' = \mu_K (1 - ih')T^s_\xi + (1 + ih')T^l_\xi . \tag{4.37}$$

Taking the total derivative along the interface of equation (4.35b), we find

$$\frac{dT^l}{d\xi} = T^l_\xi + h'T^l_\eta = (1 + ih')T^l_\xi = -\frac{1}{2}\sigma[\kappa a]' + h' . \tag{4.38}$$

This is used to replace $(1 + ih')T^l_\xi$ in (4.37),

$$(1 - ih')T^s_\xi = \frac{1}{\mu_K}\left[-[(1 - i\xi)h]' + \frac{1}{2}\sigma[\kappa a]'\right] . \tag{4.39}$$

Since

$$(1 - ih')T^s_\xi = T^s_\xi + h'T^s_\eta = \frac{dT^s}{d\xi}, \tag{4.40}$$

(4.39) can be integrated directly,

$$T^s = \frac{1}{\mu_K}\left[-(1 - i\xi)h + \frac{1}{2}\sigma\kappa a(\theta)\right] . \tag{4.41}$$

4. Extended heat transport properties

Now, everything can be put into (4.35a): On the RHS

$$\gamma_K \left(T_\eta^s - h' T_\xi^s\right) = -i\gamma_K(1 - ih') T_\xi^s,$$

(4.39) is inserted. On the LHS, $T^l - h$ is replaced using the Gibbs-Thomson condition (4.35b) and T^s is replaced using (4.41),

$$i\gamma_K \left[\frac{1}{2}\sigma\kappa a(\theta) - (1 - i\xi)h\right]' = \sqrt{(\xi^2 + (1+h)^2)(1 + h'^2)}$$
$$\times \left[\frac{1}{2}\sigma\kappa a(\theta)(1 + \mu_K) - (1 - i\xi)h\right]. \quad (4.42)$$

This is the shape equation determining the correction function $h(\xi)$. The analogon of equation (4.42) in cartesian coordinates reads

$$i\gamma_K \left[\frac{1}{2}\sigma\kappa a(\theta) - \frac{(1 - ix)\zeta}{1 + x^2}\right]' + \gamma_K \frac{(3 - x^2)x}{(1 + x^2)^3}\zeta\zeta' =$$
$$\sqrt{1 + y_s'^2}\left[\frac{1}{2}\sigma\kappa a(\theta)(1 + \mu_K) - \frac{(1 - ix)\zeta}{1 + x^2}\right] \quad (4.43)$$

with the interface position $y_s = \frac{1}{2}(1 - x^2) + \zeta(x)$. In cartesian coordinates, the curvature κ and the fourfold anisotropy function $a(\theta)$ are

$$\kappa = -\frac{y_s''}{(1 + y_s'^2)^{3/2}}, \quad (4.44)$$

$$a(\theta) = 1 - \beta\left[1 - 8\frac{y_s'^2}{(1 + y_s'^2)^2}\right]. \quad (4.45)$$

The derivation of (4.43) and (4.44)-(4.45) can be found in appendix A.3. Note, that a form of (4.43) was also derived by Fischaleck [43]. It is equation (7.17) in the cited work. However, the algebraic sign of the first term on the LHS of (4.43) is different. The reason is that Fischaleck uses a wrong form of the thermal discontinuity condition from his subsection 7.1.1 on. In addition to that, the nonlinear term

4.2. Thermal resistance at the two-phase boundary

on the LHS of (4.43) does not appear in [43], because in the cited work the expanded form of the Stefan condition was linearized in terms of ζ and its derivatives.

The stretching transformation $\xi = -i(1 - \sigma^\alpha t)$, $h(\xi) = \sigma^\alpha \phi(t)$ is used in equation (4.42). Some useful formulas can be found in the analogue transformation in the potential flow problem (see appendix A.1.4). In particular, the transformed versions of κ and $a(\theta)$ are given in equations (A.24) and (A.23) respectively. We find

$$\sqrt{(\xi^2 + (1+h)^2)(1+h'^2)} \approx \sigma^{\frac{\alpha}{2}} \sqrt{1 - \dot{\phi}^2} \sqrt{2t + 2\phi}. \tag{4.46}$$

Setting $\alpha = \frac{2}{7}$ and

$$\Lambda_K = \kappa a(\theta) \sigma^{\frac{3}{2}\alpha} = \mathcal{O}(\sigma^0), \tag{4.47}$$

the transformed version of (4.42) reads

$$\gamma_K \sigma^{-\frac{3}{7}} \left[\frac{1}{2} \dot{\Lambda}_K - \dot{\phi} - t\ddot{\phi} \right] = \sqrt{2(1 - \dot{\phi}^2)(t + \phi)} \left[\frac{1}{2}(1 + \mu_K)\Lambda_K - t\dot{\phi} \right]. \tag{4.48}$$

This third order equation can be written conveniently as a system of three first order equations with Λ_K as an additional dependent variable. We set $\phi = x_1$, $\dot{\phi} = x_2$, $\Lambda_K = x_3$, $b = \beta \sigma^{-4/7}$ and

$$p = b\gamma_K \sigma^{-\frac{3}{7}} = \frac{R_K K_s P_c \beta}{2d_0}. \tag{4.49}$$

The first order equation for $\dot{x}_2$ is derived by solving the definition (4.47) of Λ_K for $\ddot{\phi}$. The first order equation for $\dot{x}_3$ is derived by solving

4. Extended heat transport properties

(4.48) for $\dot{\Lambda}_K$,

$$\dot{x}_1 = x_2 \tag{4.50a}$$
$$\dot{x}_2 = f_K(\{x_i\}, t) \tag{4.50b}$$
$$\dot{x}_3 = g_K(\{x_i\}, t) \tag{4.50c}$$

$$f_K(\{x_i\}, t) = \frac{\sqrt{2}x_3(1-x_2)^{\frac{3}{2}}(t+x_1)^{\frac{5}{2}}(1+x_2)^{\frac{7}{2}}}{(t+x_1)^2(1+x_2)^2 - 2b(1-x_2)^2} - \frac{(1-x_2)(1+x_2)^2}{2(t+x_1)} \tag{4.51a}$$

$$g_K(\{x_i\}, t) = \frac{b}{p}\sqrt{2(1-x_2^2)(t+x_1)}\left[(1+\mu_K)x_3 - 2tx_1\right] + 2(x_1 + tx_2). \tag{4.51b}$$

A numerical calculation of the eigenvalue b from the system (4.50a)-(4.50c) was tried using the method described in section 2.3. No convergence was obtained. An implementation of the cartesian coordinate analogon of (4.48) was also tried but did not lead to success either. Perhaps the rather large value of $R_K K_s$ for ^{3}He gives rise to too many numerical inaccuracies. But the material parameters of succinonitrile with $\mu_K = 1$ and $p = 0$ as well as $p = \mathcal{O}(1)$ were also tried, neither succeeding. On the other hand, one cannot expect convergence as stable as in the convective problems for example (equations (3.30) and (3.62)), because in the thermal resistance problem, the eigenvalue b appears twice (in $f_K(\{x_i\}, t)$ and in $g_K(\{x_i\}, t)$) and not just once. The additional parameter group b/p in g_K could not be removed from the equations by a different choice of the dominant balance scaling exponent α. Hence, the classical scaling laws (1.34a) and (1.34b) for dendritic growth without convection hold even in case of a finite thermal resistance at the two-phase boundary, if a solution exists.

4.3. Anisotropic diffusion

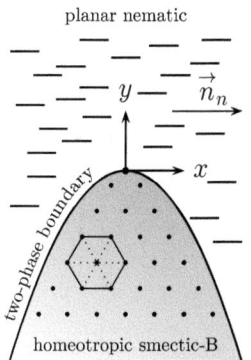

Figure 4.2.: Scheme of the liquid crystal system set-up in the focus of this section

The special role of anisotropy in microscopic solvability theory gives rise to the question, whether the presence of anisotropic diffusion alone can lead to stable solutions. This is relevant for example in liquid crystals. In this section, the selection problem of diffusion-limited dendritic growth with anisotropic heat transport is solved. The existence of a solution without anisotropy of surface tension is investigated. The focus lies on the two-dimensional case of a smectic-B liquid crystal, growing in its own undercooled nematic phase as considered by Brzsnyi et al. [17]. The orientation of the smectic-B phase is homeotropic, i.e. its director is perpendicular to the growth plane. The orientation of the nematic phase is planar, i.e. the nematic director $\vec{n}_n$ is parallel to the growth plane. The system setup is shown in figure 4.2. Thermal diffusion in the nematic phase is least efficient perpendicular to the nematic director. This is the growth direction. Experiments show that the growth velocity V takes its largest value in the direction of lowest thermal diffusivity ("inverted growth") [55, 17]. It is a counterintuitive phenomenon, because diffusion removes the latent heat from the interface and enables steady-state dendritic growth. Any half-way realistic solution should reproduce this feature.

The anisotropic heat transport properties of liquid crystals were investigated in detail by Rondelez et al. for the substance K15 [100]. They reported explicit values for D and μ, where the dimensionless parameter μ is defined below. Moreover, Buka et al. performed experiments on the nematic-smectic-B interfaces of different substances

4. Extended heat transport properties

with different crystal orientations in the smectic-B phase [28]. Their measurements yielded an estimate for the capillary length d_0.

The model equations (1.8a)-(1.10b) are written down in an adequate manner, so that they can be used as a mathematical starting point in this section. Again, the non-dimensionalizations from section 2.1 will be used. I.e. lengths are measured in units of the tip curvature radius ρ and the temperature as well as the time are rescaled appropriately. From the beginning, we look for solutions with steady-state growth velocity $V\vec{e}_y$ in a moving frame of reference attached to the dendrite. In this frame, the diffusion equation is

$$\vec{\nabla}\left(\widehat{D} \cdot \vec{\nabla} T\right) = -\rho V (\vec{e}_y \cdot \vec{\nabla}) T \qquad (4.52)$$

and the thermal diffusivity

$$\widehat{D} = \begin{pmatrix} D & 0 \\ 0 & \mu^2 D \end{pmatrix}, \quad \mu \in (0,1) \qquad (4.53)$$

has to be specified as a second-rank tensor. μ is the heat diffusion anisotropy strength. It is a dimensionless, positive number. For $\mu \to 1$, the isotropic diffusion limit is reached. The larger the absolute value of the difference $1 - \mu$, the stronger is the heat diffusion anisotropy. In the system displayed in figure 4.2, μ is less than unity. It renders the heat transport in y-direction less efficient. But theoretically, the case $\mu > 1$ might also occur to indicate that heat diffusion is faster in growth direction.

Heat transport in the smectic-B phase does not have the same anisotropy as in the nematic phase. Strictly speaking, that anisotropy should be assumed sixfold regarding the smectic-B lattice structure, in contrast to the twofold anisotropy in the nematic phase. Especially for a homeotropic orientation of the smectic-B liquid crystal, the anisotropy could be assumed much weaker than in the nematic phase or even zero. Consequently, $\widehat{D}$ cannot be assumed to be equal

4.3. Anisotropic diffusion

in the smectic-B domain and in the nematic domain, and the two phases cannot be treated symmetrically. However, in experiments the smectic-B liquid crystal remains nearly isothermal during the whole growth process [17]. Thus, thermal diffusion in the smectic-B domain can only play a marginal role. This argument suggests the applicability of a one-sided model, which significantly simplifies the problem mathematically compared to an approach with two different diffusion tensors. Hence, the field equation (4.52) will only be solved in the nematic phase.

The boundary conditions read

$$T = -\frac{d_0}{\rho}\kappa a(\theta) \qquad \text{Gibbs-Thomson,} \qquad (4.54\text{a})$$

$$\rho V \vec{e}_y \cdot \vec{n} = -\left(\widehat{D} \cdot \vec{\nabla}T\right) \cdot \vec{n} \qquad \text{Stefan condition.} \qquad (4.54\text{b})$$

T is the temperature field in the nematic phase. It is not yet measured in units of the growth Péclet number. Kinetic effects are neglected here, because they are rather fast compared with capillarity. They may decrease the growth velocity, but they play no crucial role for selection of the operating state, as long as the growth is perpendicular to the nematic director and the molecule incorporation into the smectic-B crystal involves mainly twist. There is no heat flow term $\propto \vec{\nabla}T^s$ in the Stefan condition (4.54b), because diffusion in the smectic-B domain is ignored in the one-sided model. Since convection is neglected here, the field equation (4.52) is linear, and the situation is simpler than in chapter 3. Cartesian coordinates will be used here.

4.3.1. Rescaled system and Ivantsov solution for dendritic growth with anisotropic diffusion

It turns out that the simple spatial rescaling

$$y = \mu \bar{y} \qquad (4.55)$$

of the y-axis reduces the problem to the case of isotropic diffusion. In the rescaled system, the curvature κ is replaced by the function $\bar{\kappa}$, and the interface tangent vector $\vec{t}$ and the interface normal vector $\vec{n}$ have to be substituted by their transformed versions, too. The new quantities do not have the same geometric meaning. E. g., $\bar{\kappa}$ is not the curvature in the new coordinate system. The model (4.52), (4.54a)-(4.54b) becomes

diffusion equation:
$$T_{xx} + T_{yy} + P_\mu T_y = 0 \qquad \text{in the nematic} \qquad (4.56)$$

interface conditions:
$$T = -\frac{1}{2}\sigma_\mu \bar{\kappa} a(\theta) \qquad \text{Gibbs-Thomson} \qquad (4.57a)$$
$$1 = -\left(\partial_y - y'_s \partial_x\right) T \qquad \text{Stefan condition.} \qquad (4.57b)$$

The bar over the new $\bar{y}$ has been dropped immediately. Here, T is measured in units of the effective Péclet number
$$P_\mu = \frac{\rho V}{\mu D} \qquad (4.58)$$
instead of the growth Péclet number $P_c = \mu P_\mu$, and the effective stability parameter
$$\sigma_\mu = \frac{2d_0}{\rho P_\mu} \qquad (4.59)$$
is related to its previous version by $\sigma_\mu = \mu \sigma$. The vectors $\vec{t}$ and $\vec{n}$ become
$$\vec{t} = \frac{\vec{e}_x + \mu y'_s(x)\vec{e}_y}{\sqrt{1 + \mu^2 y'^2_s(x)}} \qquad \vec{n} = \frac{\vec{e}_y - \mu y'_s(x)\vec{e}_x}{\sqrt{1 + \mu^2 y'^2_s(x)}} \qquad (4.60)$$
and $\vec{n}$ was used in the Stefan condition. These vectors are no longer tangential and perpendicular to the curve prescribed by $(x, y_s(x))$. A

4.3. Anisotropic diffusion

derivation of $\vec{t}$ and $\vec{n}$ can be found in appendix A.3.1, just the μ-factors from the rescaling of the y-axis have to be inserted to get to the forms (4.60). The Gibbs-Thomson condition (4.57a) is different because of the function

$$\bar{\kappa} = -\frac{\mu y_s''(x)}{(1 + \mu^2 y_s'^2(x))^{3/2}} \qquad (4.61)$$

and the anisotropy function $a(\theta)$. The twofold and fourfold anisotropy functions of surface tension are

$$a(\theta) = 1 + \frac{\beta_2}{2}\left(1 - \frac{2}{1 + \mu^2 y_s'^2(x)}\right), \qquad \text{twofold} \qquad (4.62a)$$

$$a(\theta) = 1 - \beta_4\left(1 - \frac{8\mu^2 y_s'^2(x)}{(1 + \mu^2 y_s'^2(x))^2}\right). \qquad \text{fourfold} \qquad (4.62b)$$

Both forms are given here, and choosing one of them is postponed to the upcoming subsections. A derivation of (4.62a)-(4.62b) can be found in appendix A.3.1, again just adding the μ-factors from the rescaling of the y-axis.

Apart from the fact, that the sought-after number P_c has been replaced by its effective "anisotropic-diffusion-specific" value, the field equation (4.56) has the same mathematical structure as in the classical case without any side effects such as convection or anisotropic heat transport. $\bar{\kappa}$ and $a(\theta)$ play no role for $\sigma_\mu = 0$ and thus, replacing P_c by P_μ in (1.17) yields an equation determining P_μ for given undercooling Δ:

$$\Delta = \sqrt{\frac{\pi P_\mu}{2}} e^{\frac{P_\mu}{2}} \operatorname{erfc}\left(\sqrt{\frac{P_\mu}{2}}\right). \qquad (4.63)$$

This equation is obtained using the far field boundary condition $\Delta = -P_\mu T|_{y\to\infty}$, and it results from the neglect of surface tension. In this case, the interface is isothermal and one can find a solution with

4. Extended heat transport properties

parabolic shape of the dendrite:

$$T^{\text{Iv}}(x,y) = -\sqrt{\frac{\pi}{2P_\mu}} e^{\frac{P_\mu}{2}} \left[\text{erfc}\left(\sqrt{\frac{P_\mu}{2}}\right) - \text{erfc}\left(\sqrt{\frac{P_\mu}{2}}\sqrt{y+\sqrt{x^2+y^2}}\right) \right]. \tag{4.64}$$

The two-phase boundary is located at

$$y_s^{\text{Iv}}(x) = \frac{1}{2}(1-x^2). \tag{4.65}$$

In the equations expanded about the case $\sigma_\mu = 0$, the limit $P_\mu \to 0$ is considered. We set $T \to T^{\text{Iv}} + T$, $y_s(x) = y_s^{\text{Iv}}(x) + \zeta(x)$. The expansion of the boundary conditions (4.57a)-(4.57b) is more difficult here than in parabolic coordinates, because the expression for the Ivantsov solution (4.64) is more complicated and it depends on both spatial coordinates. The calculations can be found in appendix A.4.1. The problem reads

diffusion equation:

$$T_{xx} + T_{yy} = 0 \qquad \text{in the nematic} \tag{4.66}$$

interface conditions:

$$T = -\frac{1}{2}\sigma_\mu \bar{\kappa} a(\theta) + \frac{\zeta(x)}{1+x^2} \qquad \text{Gibbs-Thomson} \tag{4.67a}$$

$$\left[\frac{x\zeta(x)}{1+x^2}\right]' = -\left(\partial_y - y'_s(x)\partial_x\right)T \qquad \text{Stefan condition.} \tag{4.67b}$$

for the corrections. This describes the extension to a finite anisotropic surface tension with a non-isothermal, nearly parabolic interface. From (4.66) with (4.67a)-(4.67b), an equation determining the eigenvalue σ_μ can be derived.

4.3.2. Derivation of the shape equation and its WKB solution

As explained in section 2.2, a general solution of the Laplace equation (4.66) is a superposition of a function from the kernel of the operator $\partial_x + i\partial_y$ and a function from the kernel of the operator $\partial_x - i\partial_y$:

$$T(x,y) = T_a\bigl(x + i(y - y_s)\bigr) + T_b\bigl(x - i(y - y_s)\bigr). \tag{4.68}$$

The function $\bar{\kappa}$ becomes singular at $x = -i/\mu$ for $|\zeta'(x) \ll 1|$. Writing the boundary condition (4.67a) at this point, we find that $T_a(-i/\mu) + T_b(-i/\mu)$ must also diverge. Now we consider $T\left(0, \frac{1}{2} + \frac{1}{\mu}\right)$:

$$T\left(0, \frac{1}{2} + \frac{1}{\mu}\right) = T_a(i/\mu) + T_b(-i/\mu). \tag{4.69}$$

Since $\left(0, \frac{1}{2} + \frac{1}{\mu}\right)$ is just an ordinary point in the nematic domain for $\mu > 0$, both terms on the right hand side of (4.69) must be bounded. Thus, $T_b(-i/\mu)$ cannot compensate the singularity in the Gibbs-Thomson condition and $T_a(-i/\mu)$ must diverge. This leads to $T = T_a$ close to the singularity and in the limit $P_\mu \to 0$. We find

$$T_y = iT_x, \tag{4.70}$$

because $(\partial_x + i\partial_y)T_a = 0$. This is enough information to derive a shape equation determining $\zeta(x)$. I.e. we just put (4.70) into the boundary conditions without calculating an explicit form of $T(x,y)$. We take the derivative along the interface of the Gibbs-Thomson condition (4.67a). Using the formula

$$\frac{dT}{dx} = T_x + y'_s(x)T_y = -i\Bigl(1 + iy'_s(x)\Bigr)T_y, \tag{4.71}$$

the boundary conditions become

$$-i\Bigl(1 + iy'_s(x)\Bigr)T_y = -\frac{1}{2}\sigma_\mu[\bar{\kappa}a(\theta)]' + \left[\frac{\zeta(x)}{1+x^2}\right]', \tag{4.72a}$$

4. Extended heat transport properties

$$-\left(1+iy'_s(x)\right)T_y = \left[\frac{x\zeta(x)}{1+x^2}\right]'. \tag{4.72b}$$

From (4.72a)–i(4.72b), one finds an equation free of T and its derivatives, which can be intergrated right away:

$$\frac{1}{2}\sigma_\mu \bar{\kappa} a(\theta) = \frac{\zeta(x)}{1+ix}. \tag{4.73}$$

This is the shape equation determining $\zeta(x)$. The underlying model was reduced to the case of isotropic diffusion with the exception of $\bar{\kappa}$ and $a(\theta)$, which did not have to be inserted until now. Thus, (4.73) could have also been obtained by following one of the classical solution schemes [7, 82] and neglecting some non-divergent terms. Here, it was derived in a much simpler manner by using (4.70). Fischaleck explained that this leads to the same results as the classical methods [44]. The factor $\frac{1}{2}$ in front of σ_μ does not occur in the symmetric model [88]. Equation (4.73) has the WKB solution (see appendix A.4.2)

$$\zeta(x) = B_2 \frac{\left(-\mu(1+ix)\right)^{1/4}}{\left(2\left(1+\mu^2 x^2\right)\right)^{-3/8}} \\ \times \exp\left(\sqrt{\frac{2}{\sigma_\mu}} i \int_{-i}^{x} \frac{\left(1+\mu^2 x'^2\right)^{3/4}}{\sqrt{\mu(1+ix')}} \mathrm{d}x'\right) \tag{4.74}$$

with the numerical constant B_2.

4.3.3. Derivation of the local equation and its numerical solution for dendritic growth mode selection with anisotropic diffusion

In this subsection, a local equation is derived, the solution of which is a good approximation to the function $\zeta(x)$ in the vicinity of the singular point of the problem. We start with equation (4.73) and introduce $\bar{\sigma} = \tfrac{1}{2}\sigma_\mu$. Assuming $\zeta'(x) = \mathcal{O}(\bar{\sigma}^{\alpha_\mu})$ with the sought-after scale exponent α_μ, the function $\bar{\kappa}$ from (4.61) and the functions $a(\theta)$ from (4.62a), (4.62b) become singular at $x^* = -\mathrm{i}/\mu$. The stretching transformation has to be altered:

$$x = -\frac{\mathrm{i}}{\mu}(1 - \bar{\sigma}^{\alpha_\mu} t),\qquad (4.75\mathrm{a})$$

$$\zeta(x) = \bar{\sigma}^{2\alpha_\mu}\phi(t).\qquad (4.75\mathrm{b})$$

The double scaling exponent in $\zeta(x)$ results in $\zeta'(x)$ being asymptotically small for $\bar{\sigma} \to 0$, but the second derivative $\zeta''(x)$ remains finite in this limit. A detailed execution of the transformation can be found in appendix A.4.3. $\bar{\kappa}$ becomes

$$\bar{\kappa} = \frac{\mu(1+\mu^2\ddot{\phi})}{[2(\mu^2\ddot{\phi}+t)]^{3/2}}\,\bar{\sigma}^{-\tfrac{3}{2}\alpha_\mu}\qquad (4.76)$$

and the anisotropy functions turn into

$$a(\theta) = 1 - \frac{b_2}{2(\mu^2\ddot{\phi}+t)} \qquad \text{twofold} \qquad (4.77\mathrm{a})$$

$$a(\theta) = 1 - \frac{2b_4}{(\mu^2\ddot{\phi}+t)^2} \qquad \text{fourfold} \qquad (4.77\mathrm{b})$$

with $b_2 = \beta_2/\bar{\sigma}^{\alpha_\mu}$ and $b_4 = \beta_4/\bar{\sigma}^{2\alpha_\mu}$. The local equations read

$$\frac{1+\mu^2\ddot{\phi}}{(2\tau_\mu)^{3/2}}\left(1 - \frac{b_2}{2\tau_\mu}\right) = \frac{\phi}{1+\mu} \qquad \text{twofold} \qquad (4.78\mathrm{a})$$

$$\frac{1+\mu^2\ddot{\phi}}{(2\tau_\mu)^{3/2}}\left(1 - \frac{2b_4}{\tau_\mu^2}\right) = \frac{\phi}{1+\mu} \qquad \text{fourfold} \qquad (4.78\mathrm{b})$$

for the two surface tension anisotropy types respectively with $\tau_\mu = \mu^2\ddot{\phi}+t$. One has to keep in mind that the effective stability parameter σ_μ has twice the value of $\bar{\sigma}$: $\sigma_\mu = 2\bar{\sigma} = 2\times(\beta_2/b_2)^{7/2}$. $\bar{\sigma}$ would have the

4. Extended heat transport properties

ordinary form, if the symmetric model was used. The leading terms on both hand sides of the equations are found to be balanced in their leading order of $\bar{\sigma}$, if the scale exponent is chosen to be $\alpha_\mu = \alpha = 2/7$. The eigenvalues b_2 and b_4 can be assumed to be $\mathcal{O}(1)$. For an n-fold anisotropy function of capillary effects, the scaling law

$$\sigma_n \propto \beta_n^{7/n} \tag{4.79}$$

from the case of isotropic diffusion applies also for anisotropic diffusion. The explicit formula (1.35) for the growth velocity V at small undercoolings has to be extended by an additional factor μ:

$$V = \frac{2D\Delta^4 \mu \sigma_\mu(\mu)}{d_0 \pi^2}. \tag{4.80}$$

For $\mu \to 1$, the fourfold surface tension anisotropy version (4.78b) becomes identical to the local equation of Brener and Mel'nikov [24], representing the limit of the well-known theory for dendritic growth with isotropic diffusion. In fact, equation (4.78a) can be converted into an equation having the form of the local equation from the corresponding problem with isotropic diffusion. This is achieved by rescaling:

$$t = \mu^{4/7} \left(\frac{1+\mu}{2}\right)^{2/7} \tilde{t}, \tag{4.81a}$$

$$\phi(t) = \mu^{-6/7} \left(\frac{1+\mu}{2}\right)^{4/7} \Psi(\tilde{t}). \tag{4.81b}$$

The inner equation becomes

$$\frac{1+\Psi''}{(2\tilde{\tau}_\mu)^{3/2}} \left(1 - \frac{\tilde{b}_2}{2\tilde{\tau}_\mu}\right) = \frac{\Psi}{2} \tag{4.82}$$

with $\tilde{\tau}_\mu = \tilde{t} + \Psi'$, and the nonlinear eigenvalue

$$\tilde{b}_2 = \frac{\beta_2}{\bar{\sigma}^{2/7}} \left[\frac{2}{\mu^2(1+\mu)}\right]^{2/7} \tag{4.83}$$

4.3. Anisotropic diffusion

is expressed as a function of $\bar{\sigma}$ and μ. Here, the prime denotes the derivative with respect to $\tilde{t}$. Equation (4.82) is equivalent to (4.78a) at $\mu = 1$. Only the respective eigenvalues $\tilde{b}_2$ and b_2 are defined differently. A numerical treatment of equation (4.82) yields the eigenvalue $\tilde{b}_2 = 1.6608$, which is equal to b_2 at $\mu = 1$. The fourfold surface tension anisotropy version (4.78b) was also implemented yielding $b_4 = \beta_4/\sigma^{4/7} = 0.6122$ at $\mu = 1$. This value was also found by Brener [24] and Tanveer [117]. From (4.83) one finds

$$\sigma_\mu = \frac{4}{\mu^2(1+\mu)} \left(\frac{\beta_2}{1.6608}\right)^{\frac{7}{2}} \qquad (4.84)$$

and together with $P_\mu = \rho V/(\mu D)$

$$V = \left(\frac{\beta_2}{1.6608}\right)^{\frac{7}{2}} \frac{P_\mu^2 D}{d_0} \frac{2}{\mu(1+\mu)}, \qquad (4.85a)$$

$$\rho = \left(\frac{\beta_2}{1.6608}\right)^{-\frac{7}{2}} \frac{d_0}{P_\mu} \frac{\mu^2(1+\mu)}{2}. \qquad (4.85b)$$

For $P_\mu \ll 1$, (4.63) yields $P_\mu \approx 2\Delta^2/\pi$. This is put into (4.85a),

$$V = \frac{8\Delta^4 D}{\mu(1+\mu)d_0\pi^2} \left(\frac{\beta_2}{1.6608}\right)^{\frac{7}{2}}. \qquad (4.86)$$

The same result for V could have been obtained by inserting σ_μ from (4.84) into (4.80).

In this subsection, results are shown for the substance CCH5. The twofold surface tension anisotropy is chosen here, because it is the dominant contribution for CCH5 in the configuration with a homeotropically orientated smectic-B phase and a planar orientation of the nematic phase [17]. CCH5 is a long organic molecule consisting of a pentyl chain, two cyclo-hexane rings and a nitrile group. The values of μ and D were measured for the substance K15 in [100]. But they can be assumed to be similar for CCH5, because they depend almost

4. Extended heat transport properties

only on the alkyl chain length [121], which is equal for both substances (see figure 4.3). Moreover, the value of d_0 is only a rough estimate for nematic-smectic interfaces gained from the surface tension measurements of Buka et al. [28]. The important material parameters of CCH5 are shown below.

CCH5:

- $d_0 = 5 \cdot 10^{-6}\, \mu m$
- $D = 1.25 \cdot 10^5\, \frac{\mu m^2}{s}$
- $\beta_2 = 0.06 \ldots 0.18$ [17]
- $\mu = 0.767$.

Figure 4.4 shows plots of the functions from (4.85a)-(4.85b) in the interval $\mu \in [0.1, 1]$ for different values of the dimensionless undercooling Δ and at fixed β_2. In fact, μ is tunable to a limited amount even in experiments by using substances with different alkyl chain lengths. The required values of P_μ are obtained by solving (4.63) numerically using the corresponding value of Δ. The lowest undercooling in use here is $\Delta = 0.02$ corresponding to an absolute value of about 0.23 K for CCH5. Börzsönyi et al. [17] observed the onset of the dendritic growth regime already at undercoolings $\gtrsim 0.15$ K. Thus, we are in the experimentally relevant range. At the largest undercooling $\Delta = 0.04$ used here, the effective growth Péclet number P_μ still hardly exceeds 10^{-3}. The largest possible value $\beta_2 = 0.18$ consistent with experiments for CCH5 [17] was used.

CCH5: C_5H_{11}–⬡–⬡–C≡N

K15: C_5H_{11}–⬡–⬡–C≡N

Figure 4.3.: Structure formulas of the liquid crystal molecules CCH5 and K15

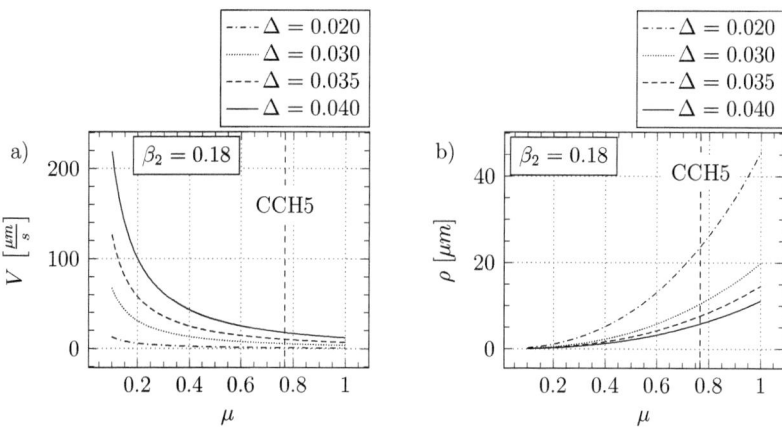

Figure 4.4.: Observable quantities; a) growth velocity V and b) tip curvature radius ρ as functions of the thermal diffusion anisotropy strength μ at different values of the dimensionless undercooling Δ. The curves are prescribed by equations (4.85a)-(4.85b). The "inverted growth" phenomenon is shown in a).

For smaller values of μ, the thermal diffusion anisotropy increases, or rather the heat transport becomes less efficient in growth direction, and the growth velocity tends to increase. This may be regarded as a full quantitative description of the "inverted growth" phenomenon observed for instance in [55]. The relevant direction for heat transport is perpendicular to the growth direction. I.e. the most heat is removed sideways (in x-direction) without increasing the temperature in front of the dendrite (in y-direction), and the growth may proceed into cooler regions. The curves in figure 4.4 are prescribed by equations (4.85a)-(4.85b), and we see that $V \to \infty$ and $\rho \to 0$ for $\mu \to 0$. I.e. in the limit, in which the latent heat is removed only laterally (in x-direction), the dendrite becomes an infinitesimal thin needle crystal growing infinitely fast.

4. Extended heat transport properties

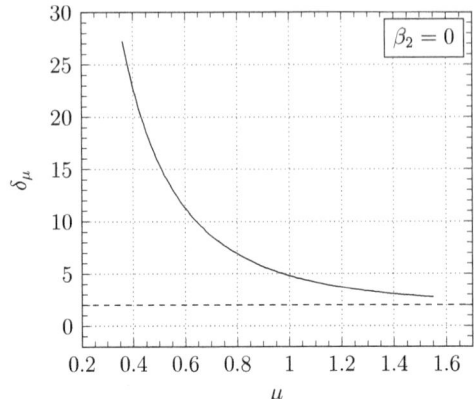

Figure 4.5.: Derivative difference δ_μ of $\dot\phi$ at the crossing point of the two lines of numerical integration as a function of the heat diffusion anisotropy strength μ (solid graph) in the case of isotropic surface tension ($\beta_2 = 0$). The approximate asymptote at $\delta_\mu = 2$ is marked by a dashed line.

As explained in section 2.3, the numerical algorithm is based on a solution of the local equation on two perpendicular lines in the complex plane. One line lies on the real axis and the other line is parallel to the imaginary axis. To obtain an analytic solution, the difference δ_μ of the derivatives $\dot\phi$ on both lines is driven to zero at the crossing point. This is achieved by adjusting b_2 or $\tilde{b}_2$, as it was the case for the results shown in figure 4.4. This procedure leads to the right eigenvalue in the case of isotropic diffusion [117]. In the case of isotropic surface tension, Kessler and Levine showed that the cusp magnitude at the dendrite tip is $\propto e^{-1/\sqrt{\sigma}}$, which never becomes zero [72]. But it would have to be zero in a physically correct situation. Thus, there is no solution with isotropic surface tension. Here, the derivative difference δ_μ cannot be observed as a function of the effective stability parameter

σ_μ for isotropic surface tension, since the choice of the scale exponent $\alpha_\mu = \frac{2}{7}$ ensured that (4.78a) is free of σ_μ for $\beta_2 = 0$. A direct solution of the outer equation (4.73) is not possible, too, at least not with the numerical procedure described in section 2.3, because the equation contains complex coefficients rendering the mandatory symmetry requirements unsustainable. But from equation (4.78a) for the case of isotropic surface tension ($b_2 = 0$), the derivative difference δ_μ can be calculated as a function of the heat diffusion anisotropy strength μ. This is shown in figure 4.5. The graph decays monotonically and it approximately approaches $\delta_\mu = 2$ for $\mu > 1.5$. The decay is non-exponential. Nevertheless, there are no zero-crossings. Thus, we find that heat diffusion anisotropy cannot stabilize the dendrite.

The numerical evidence is complemented by the fact that μ can be eliminated from equation (4.78a) by the rescaling (4.81a)-(4.81b). In the resulting equation (4.82), μ appears only in $\tilde{b}_2$, but this eigenvalue is still zero in the case of isotropic surface tension. Thus, from a mathematical point of view, it cannot be seen how the sole presence of a $\mu < 1$ could stabilize the dendrite. If selection was supposed to happen analogously to the case of finite β_2, one would need a factor, that becomes singular at $\mu = 1$. Such a factor is not present in the local equation. Obviously, (4.78a) with $b_2 = 0$ is not a new eigenvalue problem for the nonlinear eigenvalue μ, and heat diffusion anisotropy does not constitute a singular perturbation, which could break the degeneracy of Ivantsov's solution spectrum. In stark contrast to that, for finite σ_μ, the problem is drastically changed by the capillary term in the Gibbs-Thomson condition (4.57a).

An article about the results of this section has been submitted to *Advances in Condensed Matter Physics* [124]. At the time when this work is about to be finished, the article is passing through the peer review process.

4. Extended heat transport properties

4.4. Arbitrary growth Péclet numbers and asymptotic decomposition

In this subsection, the applicability of Zauderer's decomposition scheme in the case of arbitrary growth Péclet numbers P_c is investigated. In the convective problems of chapter 3, the limit $P_c \to 0$ was taken in the expansion about the Ivantsov solution. However, not all physical situations correspond to this limiting case. For instance, in rapid solidification [46, 45, 3] undercoolings of more than 200 K can be realized. In this case, P_c terms cannot be neglected.

Regarding the previous sections, experience shows that the bulk equation in the solid will not be of Laplace type, if P_c is arbitrary. As a consequence, when decomposing the solid temperature correction in eigenvectors of the first order system, one will have to use two terms, rendering the evaluation of the boundary conditions rather complicated. In order to avoid this problem, the one-sided model is used in this section. We stick to a cartesian frame,

diffusion equation:
$$T_{xx} + T_{yy} + P_c T_y = 0 \qquad \text{in the liquid} \qquad (4.87)$$

interface conditions:
$$T = -\frac{1}{2}\sigma \kappa a(\theta) \qquad \text{Gibbs-Thomson} \qquad (4.88a)$$
$$1 = -\big(\partial_y - y'_s(x)\partial_x\big)T \qquad \text{Stefan condition.} \qquad (4.88b)$$

Here, the equations are already written in non-dimensional form in the moving frame. T is the temperature in the liquid and it is measured in units of P_c. In this one-sided model, diffusion in the solid is ignored completely. For this reason, the Stefan condition has only the heat balance term belonging to the liquid phase.

4.4.1. Transformation of the temperature field

For vanishing surface tension, the interface is located at $y_s^{\text{Iv}}(x) = \frac{1}{2}(1-x^2)$, and the Ivantsov condition resulting from equation (4.87) is given by (1.17). Again, the solution is expanded about this case. For the temperature correction, equation (4.87) applies due to its linearity property. With the two-phase boundary shifted to $y_s(x) = y_s^{\text{Iv}}(x) + \zeta(x)$, the boundary conditions read

$$T = -\frac{1}{2}\sigma\kappa a(\theta) + \frac{\zeta(x)}{1+x^2} \qquad \text{Gibbs-Thomson} \tag{4.89a}$$

$$(\partial_y - y_s'(x)\partial_x)T = -\left[\frac{x\zeta(x)}{1+x^2}\right]' - P_c\frac{\zeta(x)}{1+x^2} \qquad \text{Stefan condition} \tag{4.89b}$$

and they can be obtained by using appendix A.4.1. There, the equations are given in the framework of the symmetric model for a nematic growing in its own smectic-B phase. Nevertheless, it can be used, replacing P_μ by P_c and ignoring T^s and its derivatives because of the one-sided model. A term of the order $\mathcal{O}(\zeta\zeta')$ was neglected in (4.89b).

The transformation

$$T = \bar{T}e^{-\frac{P_c}{2}y} \tag{4.90}$$

is introduced, yielding

$$T_y = e^{-\frac{P_c}{2}y}\left(\bar{T}_y - \frac{P_c}{2}\bar{T}\right) \tag{4.91a}$$

$$T_{yy} = e^{-\frac{P_c}{2}y}\left(\bar{T}_{yy} - P_c\bar{T}_y + \frac{P_c^2}{4}\bar{T}\right) \tag{4.91b}$$

for the derivatives of T with respect to y. The diffusion equation for the temperature field correction reads

$$T_{xx} + T_{yy} - \frac{P_c^2}{4}T = 0 \tag{4.92}$$

4. Extended heat transport properties

where the bar over $\bar{T}$ has been dropped right away. In the boundary conditions, an additional term proportional to T appears in the Stefan condition as well as in the differentiated form (along the interface) of the Gibbs-Thomson condition. This term can immediately be substituted using (4.89a),

$$T_x + y'_s T_y = -\left[\left[\frac{1}{2}\sigma\kappa a(\theta) - \frac{\zeta(x)}{1+x^2}\right]'\right.$$
$$\left. + \frac{P_c}{2} y'_s \left(\frac{1}{2}\sigma\kappa a(\theta) - \frac{\zeta(x)}{1+x^2}\right)\right] e^{\frac{P_c}{2} y_s(x)} \qquad \text{Gibbs-Thomson,}$$

(4.93a)

$$T_y - y'_s T_x = -\left[\left[\frac{x\zeta(x)}{1+x^2}\right]'\right.$$
$$\left. + \frac{P_c}{2}\left(\frac{1}{2}\sigma\kappa a(\theta) + \frac{\zeta(x)}{1+x^2}\right)\right] e^{\frac{P_c}{2} y_s(x)} \qquad \text{Stefan condition.}$$

(4.93b)

4.4.2. Continuation to the complex plane and asymptotic decomposition (arbitrary Péclet numbers)

We would like to decompose equation (4.92) with interface conditions (4.93a) and (4.93b) à la Zauderer. In contrast to section 2.2, we employ the three-component variable $\vec{\vartheta} = (T_x, T_y, T)^{\mathrm{T}}$ instead of a two-component vector to write (4.92) as a first order system. Algebraically, the calculations have several aspects in common with asymptotic decomposition in the case of nonlinear diffusion (subsection 4.1.2). I.e. we can use the matrix E from (4.14), its eigenvectors $\vec{r}_{3,4,5}$ from (4.17)

4.4. Arbitrary growth Péclet numbers and asymptotic decomposition

and the corresponding projection operators $P_{3,4,5}$ from (4.20). In addition to that, we define

$$G = \begin{pmatrix} 0 & 0 & P_c^2/4 \\ 0 & 0 & 0 \\ -1 & -1 & 0 \end{pmatrix} \quad (4.94)$$

and find the first order system

$$\vec{\vartheta}_x + E\vec{\vartheta}_y + G\vec{\vartheta} = 0 \quad (4.95)$$

arising from (4.92). Using the ansatz

$$\vec{\vartheta} = M\vec{r}_3 + \varepsilon N\vec{r}_4 + Q\vec{r}_5 \quad (4.96)$$

and the scaling transformation $(x, y) \to (\varepsilon x, \varepsilon y)$, (4.95) becomes

$$(M_x + iM_y)\vec{r}_3 + \varepsilon(N_x - iN_y)\vec{r}_4 + (Q_x + Q_y)\vec{r}_5$$
$$+ \varepsilon GM\vec{r}_3 + \varepsilon GQ\vec{r}_5 = \mathcal{O}\left(\varepsilon^2\right). \quad (4.97)$$

We use the formulas

$$P_3 G \vec{r}_3 = 0 \qquad P_4 G \vec{r}_3 = 0 \qquad P_5 G \vec{r}_3 = (i-1)\vec{r}_5 \quad (4.98\text{a})$$

$$P_3 G \vec{r}_5 = \frac{i}{8} P_c^2 \vec{r}_3 \qquad P_4 G \vec{r}_5 = -\frac{i}{8} P_c^2 \vec{r}_4 \qquad P_5 G \vec{r}_5 = 0. \quad (4.98\text{b})$$

They are used to project (4.97) onto the invariant subspaces of E:

$$M_x + iM_y = -\frac{P_c^2}{8} iQ \quad (4.99\text{a})$$

$$N_x - iN_y = \frac{P_c^2}{8} iQ \quad (4.99\text{b})$$

$$Q_x + Q_y = (1-i)M. \quad (4.99\text{c})$$

Here, we have returned to the original scale after figuring out the principal part of the equation. The characteristic coordinates are

for (4.99a):

4. Extended heat transport properties

$$s = x + i\left(1 - \sqrt{2(y - ix)}\right) \quad \tau = -i\left(1 - \sqrt{2(y - ix)}\right) \quad (4.100a)$$

$$x = \tau + s \quad y = is - \frac{1}{2}(\tau^2 - 1) \quad (4.100b)$$

for (4.99b):

$$\bar{s} = x - i\left(1 - \sqrt{2(y + ix)}\right) \quad \bar{\tau} = i\left(1 - \sqrt{2(y + ix)}\right) \quad (4.100c)$$

$$x = \bar{\tau} + \bar{s} \quad y = -i\bar{s} - \frac{1}{2}(\bar{\tau}^2 - 1) \quad (4.100d)$$

for (4.99c):

$$\tilde{s} = x + 1 - \sqrt{2(y - x)} \quad \tilde{\tau} = -1 + \sqrt{2(y - x)} \quad (4.100e)$$

$$x = \tilde{\tau} + \tilde{s} \quad y = \tilde{s} - \frac{1}{2}(\tilde{\tau}^2 - 1). \quad (4.100f)$$

They are chosen appropriately, in a manner so that the interface is at $s, \bar{s}, \tilde{s} = 0$ and $\tau, \bar{\tau}, \tilde{\tau} = x \Leftrightarrow s, \bar{s}, \tilde{s} = 0$. We consider only contributions up to the first order in P_c. So the terms on the right hand sides of (4.99a) and (4.99b) are dropped, and the system of differential equations decouples:

$$M_s = 0 \quad (4.101a)$$
$$N_{\bar{s}} = 0 \quad (4.101b)$$
$$Q_{\tilde{s}} = (1 - i)M. \quad (4.101c)$$

Now we go back to the boundary conditions (4.93a), (4.93b). We insert $T_x = -i(M - N)$, $T_y = M + N$:

$$-i\left(1 + iy'_s(x)\right)M + i\left(1 - iy'_s(x)\right)N = -\left[\frac{1}{2}\sigma\kappa a(\theta) - \frac{\zeta(x)}{1 + x^2}\right]'$$

$$+ \frac{P_c}{2}y'_s(x)\left(\frac{1}{2}\sigma\kappa a(\theta) - \frac{\zeta(x)}{1 + x^2}\right)e^{\frac{P_c}{2}y_s(x)}, \quad (4.102a)$$

4.4. Arbitrary growth Péclet numbers and asymptotic decomposition

$$i\Big(1 + iy'_s(x)\Big)M + i\Big(1 - iy'_s(x)\Big)N =$$
$$- \left[\left[\frac{x\zeta(x)}{1+x^2}\right]' + \frac{P_c}{2}\left(\frac{1}{2}\sigma\kappa a(\theta) + \frac{\zeta(x)}{1+x^2}\right)\right]e^{\frac{P_c}{2}y_s(x)}. \quad (4.102b)$$

We perform i(4.102a) + (4.102b) and −i(4.102a) + (4.102b) in order to find the following boundary conditions:

$$M(s=0,\tau) = -\frac{i}{2(1+iy'_s(\tau))}e^{\frac{P_c}{2}y_s(\tau)}\left[\left[\frac{1}{2}\sigma\kappa a(\theta) - \frac{(1+i\tau)}{1+\tau^2}\zeta(\tau)\right]'\right.$$
$$\left.-\frac{i}{2}P_c\left[\Big(1+iy'_s(\tau)\Big)\frac{1}{2}\sigma\kappa a(\theta) + \Big(1-iy'_s(\tau)\Big)\frac{\zeta(\tau)}{1+\tau^2}\right]\right], \quad (4.103a)$$

$$N(\bar{s}=0,\bar{\tau}) = \frac{i}{2(1-iy'_s(\bar{\tau}))}e^{\frac{P_c}{2}y_s(\bar{\tau})}\left[\left[\frac{1}{2}\sigma\kappa a(\theta) - \frac{(1-i\bar{\tau})}{1+\bar{\tau}^2}\zeta(\bar{\tau})\right]'\right.$$
$$\left.+\frac{i}{2}P_c\left[\Big(1-iy'_s(\bar{\tau})\Big)\frac{1}{2}\sigma\kappa a(\theta) + \Big(1+iy'_s(\bar{\tau})\Big)\frac{\zeta(\bar{\tau})}{1+\bar{\tau}^2}\right]\right]. \quad (4.103b)$$

From (4.101a) and (4.101b), we see that M and N must be completely independent of s and $\bar{s}$ respectively. Thus, we can read the solutions from the boundary conditions:

$$M(\tau) = -\frac{i}{2(1+iy'_s(\tau))}e^{\frac{P_c}{2}y_s(\tau)}\left[\left[\frac{1}{2}\sigma\kappa a(\theta) - \frac{(1+i\tau)}{1+\tau^2}\zeta(\tau)\right]'\right.$$
$$\left.-\frac{i}{2}P_c\left[\Big(1+iy'_s(\tau)\Big)\frac{1}{2}\sigma\kappa a(\theta) + \Big(1-iy'_s(\tau)\Big)\frac{\zeta(\tau)}{1+\tau^2}\right]\right], \quad (4.104a)$$

$$N(\bar{\tau}) = \frac{i}{2(1-iy'_s(\bar{\tau}))}e^{\frac{P_c}{2}y_s(\bar{\tau})}\left[\left[\frac{1}{2}\sigma\kappa a(\theta) - \frac{(1-i\bar{\tau})}{1+\bar{\tau}^2}\zeta(\bar{\tau})\right]'\right.$$
$$\left.+\frac{i}{2}P_c\left[\Big(1-iy'_s(\bar{\tau})\Big)\frac{1}{2}\sigma\kappa a(\theta) + \Big(1+iy'_s(\bar{\tau})\Big)\frac{\zeta(\bar{\tau})}{1+\bar{\tau}^2}\right]\right]. \quad (4.104b)$$

4. Extended heat transport properties

Since

$$Q(\tilde{s}=0,\tilde{\tau}) = -\left[\frac{1}{2}\sigma\kappa a(\theta) - \frac{\zeta(\tilde{\tau})}{1+\tilde{\tau}^2}\right] e^{\frac{P_c}{2}y_s(\tilde{\tau})}, \qquad (4.105)$$

as can be deduced from (4.89a) with $T = Q$, we find the solution for equation (4.101c):

$$\begin{aligned} Q(\tilde{s},\tilde{\tau}) = & \ (1-\mathrm{i})\int_0^{\tilde{s}} M\!\left(\tau(\omega,\tilde{\tau})\right) d\omega \\ & - \left[\frac{1}{2}\sigma\kappa a(\theta) - \frac{\zeta(\tilde{\tau})}{1+\tilde{\tau}^2}\right] e^{\frac{P_c}{2}y_s(\tilde{\tau})}. \end{aligned} \qquad (4.106)$$

In order to fulfill the far field boundary condition, Q must vanish in the limit $\tilde{s} \to \infty$,

$$\frac{1}{2}\sigma\kappa a(\theta) = \frac{\zeta(\tilde{\tau})}{1+\tilde{\tau}^2} + (1-\mathrm{i}) e^{-\frac{P_c}{2}y_s(\tilde{\tau})} \int_0^{\infty} M\!\left(\tau(\omega,\tilde{\tau})\right) d\omega. \qquad (4.107)$$

It would be very convenient, if $\tau(\omega,\tilde{\tau})$ in the integrand in (4.107) were a simple expression, say x':

$$\tau = -\mathrm{i}\left(1 - \sqrt{2(y-\mathrm{i}x)}\right) \stackrel{!}{=} x'$$
$$\Rightarrow (1-\mathrm{i}x') = \sqrt{2(y-\mathrm{i}x)}$$
$$(1-\mathrm{i}x')^2 = 2y - 2\mathrm{i}x = 2\omega(1-\mathrm{i}) + (1-\mathrm{i}\tilde{\tau})^2.$$

In the last line, (4.100f) was inserted with $\tilde{s}$ replaced by ω. Thus, we use

$$\omega = \frac{1}{2}\frac{(1-\mathrm{i}x')^2 - (1-\mathrm{i}\tilde{\tau})^2}{1-\mathrm{i}}, \quad d\omega = -\mathrm{i}\frac{1-\mathrm{i}x'}{1-\mathrm{i}} dx' \qquad (4.108)$$

as a substitution in the integral in (4.107). The denominator of the differential $d\omega$ cancels the factor in front of the integral. In addition

4.4. Arbitrary growth Péclet numbers and asymptotic decomposition

to that, we rename $\tilde{\tau} = x$,

$$\frac{1}{2}\sigma\kappa a(\theta) = \frac{\zeta(x)}{1+x^2} + \mathrm{i}\,\mathrm{e}^{-\frac{P_c}{2}y_s(x)} \int\limits_{\mathrm{i}\infty}^{x} (1-\mathrm{i}x')M(x')\,\mathrm{d}x'. \qquad (4.109)$$

The next step is to simplify this equation a bit by approximating $y_s(x) \approx y_s^{\mathrm{lv}}(x)$. Doing this, only asymptotically small terms drop out. They are $\mathcal{O}(\zeta^2)$, $\mathcal{O}(\zeta\zeta')$, $\mathcal{O}(\zeta'^2)$, $\mathcal{O}(\zeta\sigma)$ or $\mathcal{O}(\zeta'\sigma)$. Anyway, they can be neglected,

$$\begin{aligned}\frac{1}{2}\sigma\kappa a(\theta) &= \frac{\zeta(x)}{1+x^2} + \frac{1}{2}\int\limits_{\mathrm{i}\infty}^{x} \mathrm{e}^{\frac{P_c}{4}(x^2-x'^2)}\left[\left[\frac{1}{2}\sigma\kappa a(\theta) - \frac{(1+\mathrm{i}x')}{1+x'^2}\zeta(x')\right]'\right.\\ &\quad\left. - \frac{\mathrm{i}}{2}P_c\left((1-\mathrm{i}x')\frac{1}{2}\sigma\kappa a(\theta) + (1+\mathrm{i}x')\frac{\zeta(x')}{1+x'^2}\right)\right]\mathrm{d}x' \\ &= \frac{\zeta(x)}{1+x^2} + \frac{1}{2}\left[\mathrm{e}^{\frac{P_c}{4}(x^2-x'^2)}\left(\frac{1}{2}\sigma\kappa a(\theta) - \frac{(1+\mathrm{i}x')}{1+x'^2}\zeta(x')\right)\right]_{\mathrm{i}\infty}^{x} \\ &\quad + \frac{1}{2}\int\limits_{\mathrm{i}\infty}^{x} \mathrm{e}^{\frac{P_c}{4}(x^2-x'^2)}\left[\frac{P_c}{2}x'\left(\frac{1}{2}\sigma\kappa a(\theta) - \frac{(1+\mathrm{i}x')}{1+x'^2}\zeta(x')\right)\right.\\ &\quad\left. - \frac{\mathrm{i}}{2}P_c\left((1-\mathrm{i}x')\frac{1}{2}\sigma\kappa a(\theta) + (1+\mathrm{i}x')\frac{\zeta(x')}{1+x'^2}\right)\right]\mathrm{d}x' \\ &= \frac{1}{4}\sigma\kappa a(\theta) + \frac{1}{2}\frac{(1-\mathrm{i}x)}{1+x^2}\zeta(x) - \frac{\mathrm{i}}{4}P_c\int\limits_{\mathrm{i}\infty}^{x} \mathrm{e}^{\frac{P_c}{4}(x^2-x'^2)} \\ &\quad \times \left[\frac{1}{2}\sigma\kappa a(\theta) + \frac{\zeta(x')}{1+x'^2}\underbrace{(1+\mathrm{i}x')(1-\mathrm{i}x')}_{=1+x'^2}\right]\mathrm{d}x'.\end{aligned}$$

4. Extended heat transport properties

Here, integration by parts was used. This leads to the final equation of this section:

$$\sigma \kappa a(\theta) = 2\frac{(1-\mathrm{i}x)}{1+x^2}\zeta(x) \\ - \frac{\mathrm{i}}{2}P_c \int_{\mathrm{i}\infty}^{x} e^{\frac{P_c}{4}(x^2-x'^2)}\Big(\sigma \kappa a(\theta) + 2\zeta(x')\Big)\,\mathrm{d}x'\,. \tag{4.110}$$

It is an equation determining $\zeta(x)$ in cartesian coordinates and selecting the eigenvalue σ for arbitrary growth Péclet numbers P_c. In the limit $P_c \to 0$, we retrieve the well-known result of Misbah [88] for the one-sided model, where the value of the stability parameter is twice the one from the symmetric model. One can see that under a formal stretching transformation $x = -\mathrm{i}(1-\sigma^\alpha t)$, the integral on the right hand side of (4.110) is of the order $\mathcal{O}(\sigma^{3\alpha})$. Thus, it scales as the flow terms in chapter 3. This makes sense indeed, because the P_c-term in the diffusion equation (4.87) arises from the moving frame of reference. In this frame, a uniform flow is present in the liquid, even if convection is neglected in the laboratory system.

4.5. Kinetic effects

In the preceding chapters, several potential reasons for discrepancies between experiments and the theory in this work were discussed:

- flow field approximation inaccuracies,

- anisotropy strength measurement inaccuracies,

- chemical impurities,

- kinetic effects of atomic transfer at the two-phase boundary.

The latter are investigated in more detail and quantified in this section.

Kinetic effects of atomic adsorption to the solid surface further decrease the interface temperature, which now depends on the growth velocity. The full boundary condition (1.9b) with all terms has to be taken into account. It was examined in detail by Brener [22]. He finds that at low undercoolings, the dendrite grows in the direction of maximal surface tension as in the ordinary case, whereas at high undercoolings, the growth velocity strongly increases and the direction of maximal kinetic effects is favoured for dendritic growth. Regarding the method of the current work, an additional term occurs in the local equation. The anisotropy function of kinetic effects is assumed to be fourfold and has the same structure as the anisotropy function of capillary effects (e. g. see eq. (4.45)):

$$\bar{\beta}(\theta) = \beta_0 \big(1 - \beta_{kin} \cos(4(\theta - \theta_{kin}))\big). \tag{4.111}$$

β_{kin} is the strength of the fourfold anisotropy of kinetic effects, and β_0 is the strength of kinetic effects averaged over all angles θ. (β_0 is the kinetic equivalent of the mean capillary length d_0.) θ_{kin} is the angle between the growth direction and the direction of minimal $\bar{\beta}(\theta)$. Regarding equation (1.9b), all that has to be done is to apply the stretching transformation $x = -i(1 - \sigma^{2/7}t)$, $\zeta(x) = \sigma^{4/7}\phi(t)$ to the kinetic correction term $\sigma_{kin} \vec{e}_y \cdot \vec{n} \, \bar{\beta}(\theta)$. I.e. we only need

$$\vec{e}_y \cdot \vec{n} = \frac{1}{\sqrt{1 + y_s'^2(x)}} = \frac{1}{\sqrt{1 + (\zeta'(x) - x)^2}} \approx \frac{\sigma^{-1/7}}{\sqrt{2\tau_2}}. \tag{4.112}$$

The interface unit normal vector $\vec{n}$ was taken from (A.97). Obviously, the kinetic term can simply be brought forward into the inner equation, since solely the Gibbs-Thomson condition has changed, whereas the rest of the model remains the same. Equation (4.78b) with $\mu = 1$ can be used, adding the kinetic term. We find another form of the

4. Extended heat transport properties

local equation for $P_c \to 0$:

$$\frac{1+\ddot{\phi}}{(2\tau_2)^{3/2}}\left(1 - \frac{2\beta\sigma^{-4/7}}{\tau_2^2}\right) + \frac{\sigma^{-5/7}\sigma_{kin}}{\sqrt{2\tau_2}}\left(1 - \frac{2\beta_{kin}\sigma^{-4/7}}{\tau_2^2}\right) = \frac{\phi}{2}. \quad (4.113)$$

Here we have $\tau_2 = t + \dot{\phi}$ again, and

$$\sigma_{kin} = \frac{2\beta_0 V}{P_c} \quad (4.114)$$

is the stability parameter of kinetic effects resulting from the non-dimensionalizations. It is a quantitative measure for the strength of the kinetic effects. Equation (4.113) results from a calculation in cartesian coordinates, and it is written down for the case in which the directions of maximum surface tension and maximum kinetic effects coincide ($\theta_{kin} = 0$). Together with the Ivantsov condition (1.17), equation (4.113) contains enough information to select the operating state of dendritic growth with anisotropic capillary effects *and* anisotropic kinetic effects. For its derivation, $\sigma^{2/7}$ was used as the small parameter for the singular perturbation expansion and the stretching transformation. The anisotropy function of kinetic effects has the same singularity as the anisotropy function of capillary effects and the curvature. The divergence of these functions for $\sigma \to 0$ is crucial. There is a discrete spectrum (σ, σ_{kin}), only one data pair out of which corresponds to the stable solution [22]. For numerical investigation, either an algorithm determining two eigenvalues simultaneously has to be made up, or the ratio σ_{kin}/σ has to be set to some fixed value. Interesting simplifications to (4.113) are the cases of

1. isotropic kinetic effects ($\beta_{kin} = 0$),

2. isotropic surface tension ($\beta = 0$).

4.5.1. Limit of isotropic kinetic effects and anisotropic surface tension

In case 1. the corresponding local equation reads

$$\frac{1+\ddot{\phi}}{(2\tau_2)^{3/2}}\left[1-\frac{2b}{\tau_2^2}\right]+\frac{k^{5/7}}{\sqrt{2\tau_2}}=\frac{\phi}{2} \qquad (4.115)$$

with $b = \beta\sigma^{-4/7}$ and the ratio

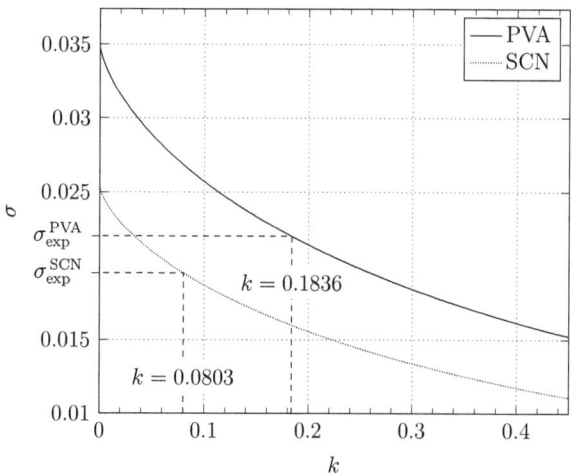

Figure 4.6.: Kinetic corrections to the stability parameter σ as a function of the ratio $k = \sigma_{kin}^{7/5}/\sigma$: Results are determined numerically from equation (4.115) with anisotropy of surface tension and without kinetic anisotropy. σ_{kin} is a quantitative measure for the strength of the isotropic kinetic effects. At $k = 0.0803$, the experimental eigenvalue $\sigma = 0.0195$ [64] for succinonitrile (SCN) is reached. At $k = 0.1836$ the experimental eigenvalue $\sigma = 0.022$ [52] for pivalic acid (PVA) is reached.

4. Extended heat transport properties

$$k = \frac{\sigma_{kin}^{7/5}}{\sigma}. \qquad (4.116)$$

This can be obtained directly by setting $\beta_{kin} = 0$ in equation (4.113). Assuming $k = \mathcal{O}(1)$, the stability parameters for pivalic acid (PVA) and succinonitrile (SCN) are calculated as functions of k from equation (4.115). This is shown in figure 4.6. k is varied in the interval $[0, 0.45]$, and σ is a monotonically decreasing function. At $k = 0.0803$, the experimental value of the stability parameter $\sigma = 0.0195$ measured by Huang and Glicksman for SCN [64] is successfully reproduced by the numerical calculation. At $k = 0.1836$, the same statement applies for PVA ($\sigma = 0.022$ [52]). However, this does *not* mean that the strengths of the kinetic effects for SCN and PVA have been determined here. A solution with anisotropy of kinetic effects would contain more convincing information and would be more desirable (see case 2. in the next subsection).

4.5.2. Limit of isotropic surface tension and anisotropic kinetic effects

In case 2. (anisotropic kinetic effects, $\beta_{kin} > 0$), a stable solution can be selected, even though the anisotropy of surface tension is neglected ($\beta = 0$) [22]. $\sigma_{kin}^{4/5}$ is used for the stretching transformation. The corresponding inner equation in cartesian coordinates becomes

$$\frac{1}{k}\frac{1+\ddot{\phi}}{(2\tau_2)^{3/2}} + \frac{1}{\sqrt{2\tau_2}}\left[1 - \frac{2b_{kin}}{\tau_2^2}\right] = \frac{\phi}{2} \qquad (4.117)$$

with $b_{kin} = \beta_{kin}/\sigma_{kin}^{4/5}$. Again, (4.117) is written down for the case, in which the directions of maximum surface tension and maximum kinetic effects coincide ($\theta_{kin} = 0$). The equation was solved numerically and the eigenvalue b_{kin} was obtained as a function of k. This is shown in figure 4.7a. k was varied in the interval $[0.2, 3]$. For $k < 1$, where

4.5. Kinetic effects

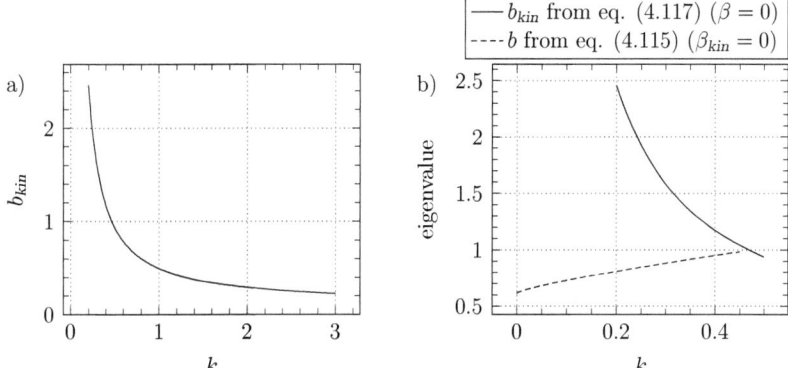

Figure 4.7.: a) selected eigenvalue $b_{kin} = \beta_{kin}/\sigma_{kin}^{4/5}$ as a function of the ratio $k = \sigma_{kin}^{7/5}/\sigma$ calculated numerically with isotropic surface tension and anisotropic kinetic effects from equation (4.117), b) b_{kin} calculated numerically from (4.117) (solid graph) in comparison to the selected eigenvalue $b = \beta/\sigma^{4/7}$ calculated numerically with isotropic kinetic effects and anisotropic surface tension from equation (4.115) (densely dashed graph)

the anisotropic kinetic effects are weaker than the isotropic capillary effects, the kinetic eigenvalue b_{kin} decreases rapidly for increasing k. The stability parameters can be calculated at any point (k, b_{kin}) of the plotted curve for given β_{kin} (and for $\beta = 0$):

$$\sigma_{kin} = \left(\frac{\beta_{kin}}{b_{kin}}\right)^{\frac{5}{4}}, \qquad (4.118a)$$

$$\sigma = \frac{1}{k}\left(\frac{\beta_{kin}}{b_{kin}}\right)^{\frac{7}{4}}. \qquad (4.118b)$$

However, experimental values for β_{kin} were not found at all during this work's literature enquiry. Hence, σ for SCN and PVA cannot be

4. Extended heat transport properties

calculated here. The only substance, for which an experimental value of β_0 was found, is Nickel ($\beta_0 = 0.002 \frac{s}{m}$, $\beta = 0.27$ [45, 35]).

In figure 4.7b, the eigenvalues from both limits $\beta = 0$ (isotropic surface tension, solid line) and $\beta_{kin} = 0$ (isotropic kinetic effects, densely dashed line) are compared. So the two curves result from the solution of different problems. Again, the data for b_{kin} was calculated from equation (4.117). The data for b as a function of k was obtained by solving equation (4.115) numerically. In contrast to b_{kin}, the eigenvalue b governed by anisotropic capillary effects increases as a function of k. Here, the stability parameters can be calculated at any point (k, b) of the dashed curve for given β using the relations

$$\sigma = (\beta/b)^{\frac{7}{4}} \tag{4.119a}$$
$$\sigma_{kin} = k^{5/7} (\beta/b)^{\frac{5}{4}} \tag{4.119b}$$

in the case of isotropic kinetic effects ($\beta_{kin} = 0$).

5. Conclusion

When a crystal germ nucleates in an undercooled liquid, a solid starts to grow. The temperature of the liquid is below the freezing point. Thus, the system is in a non-equilibrium state (but one may usually assume local equilibrium at the interface and metastable equilibrium in front of it). A planar solidification front is unstable against the *Mullins-Sekerka instability* [90]. Instead, the two-phase boundary takes a dendritic shape. The number of dendrites growing from a single nucleus depends on the symmetry type of the crystal lattice and on the size of the crystal. The bulk material is strongly affected by the properties of these microstructures, for instance by their size. The tilt and interlocking of the dendritic branches increases the robustness of a sample. For this reason, understanding and eventually controling the growth process is a desirable achievement.

Some transport mechanism has to be included into any model describing the problem, because latent heat is released at the interface during the solidification process. If the heat was not removed, the temperature in the vicinity of the interface would increase up to the melting temperature, and the crystal would cease to grow, taking its equilibrium shape. Hence, the system has to be cooled externally. In the simplest models, the heat is transported solely by thermal diffusion. This case of "classic" dendritic growth was solved numerically by Kessler and Levine [72]. In the same year, a rigorous analytical solution was presented by Ben Amar and Pomeau [7]. The authors built their solution on the older work of Ivantsov, who neglected surface tension and found a continuous family of solutions corresponding

5. Conclusion

to parabolic shapes of the dendrites[66]. However, these solutions are all structurally unstable, due to the lack of the stabilizing effect of surface tension. Capillary effects, the anisotropy of which is crucial, were found to introduce the sought-after stabilization mechanism. In the framework of singular perturbation theory starting from Ivantsov's solution, the solution spectrum is converted into a discrete set. Therefore, it is also called a "selection problem". But only one member out of the discrete set is stable. Unfortunately, for the description of most practical situations, the models need to be extended. This is where this work joins in. It represents progress in the theoretical investigation of various additional effects, which are not considered in classical dendritic growth.

First, convection is taken into account. Not only does it constitute an additional transport mechanism, but it drastically complicates the mathematical structure of the problem. The number of bulk equations is increased and they become nonlinear, rendering them non-accessible to Green's functions techniques. Approaches to the convective problem have been made in the past [19], [108], [107], [85], but they were unsatisfactory due to their simplifications. The key to the rather sophisticated problem was found by Fischaleck [44]. He used asymptotic decomposition [125] to manipulate the bulk equations directly instead of dealing with the boundary integral equation. The investigation of the scope of this method is one purpose of this work. It is used to derive analytical solutions to the dendritic growth problem with convection, and the corresponding results are published in [122, 123]. An analytical expression for the flow velocity is required in the case without surface tension. The method is predominantly analytical, except the last step, for which a C program was written. A potential flow and an Oseen flow approximation are employed. A forced flow velocity is introduced into the problem as an extra control parameter (additional to the undercooling). It is found that this parameter has only a marginal influence on the selected eigenvalue in the considered flow velocity range, especially in the Oseen flow case. However, the presence of a flow strongly changes the scaling laws in the system. For an ammonium bromide solution, the results of this work agree well with experiments, whereas there is no good agreement with flow experiments for the substances pivalic acid (PVA) and succinonitrile (SCN).

Further effects investigated in this work are for instance nonlinear diffusion and a finite thermal resistance at the two-phase boundary (*Kapitza effect*). In the first case of these two, the heat conductivity depends on the temperature, and a coupled system of differential equations for the temperature field correction is derived. In the case of Kapitza effect, an eigenvalue equation determining the shape correction of the parabolic dendrite is derived in parabolic and cartesian coordinates.

Finally, this work exhibits a rigorous solution of the selection problem of dendritic growth in systems with anisotropic diffusion. A full quantitative description of the "inverted growth" phenomenon is given (faster growth in the direction of less efficient heat transport). One might suspect that the sole presence of diffusion anisotropy can stabilize the pattern. It is found that it cannot, and the scaling laws remain the same as in the classical case. Most of the presented results are enabled by the combination of singular perturbation theory and asymptotic decomposition. It is a powerful tool, largely enhancing the applicability of the hitherto existing methods.

Appendix A.

Auxiliary calculations

A.1. Potential flow

A.1.1. Setup of the model equations

Field equations

We start with (1.8a)-(1.8b). After Galilean transformation ($\vec{r} \to \vec{r} + Vt\vec{e}_y$, $\vec{w} \to \vec{w} + V\vec{e}_y$) we find

$$\frac{\partial T^{s,l}}{\partial t} \to -V\vec{e}_y \cdot \vec{\nabla} T^{s,l}. \tag{A.1}$$

The diffusion equation (1.8b) in the liquid is invariant under that operation. The time derivative and the term in $\vec{w}$ containing V compensate each other. Now, the non-dimensionalizations from section 2.1 are applied:

$$(1.8b) \to \Delta T^l = (\vec{w} \cdot \vec{\nabla}) T^l, \tag{A.2a}$$
$$(1.8a) \to \Delta T^s = -P_c \vec{e}_y \cdot \vec{\nabla} T^s. \tag{A.2b}$$

Using the parabolic coordinate expressions (2.3a) and (2.3b) for $\vec{\nabla}$ and Δ respectively, as well as the definition (2.11) of ψ, we end up with the field equations (3.1a)-(3.1b). For the field equation (3.1c) we

Appendix A. Auxiliary calculations

just note

$$0 \overset{!}{=} \vec{\nabla} \times \vec{w} = \vec{\nabla} \times \left(\vec{\nabla} \times (\psi \vec{e}_z) \right) = \vec{\nabla} \left(\vec{\nabla} \cdot (\psi \vec{e}_z) \right) - \Delta(\psi \vec{e}_z)$$

$$= \vec{\nabla} \underbrace{\frac{\partial \psi}{\partial z}}_{=0} - (\Delta \psi) \vec{e}_z$$

$$\Rightarrow \Delta \psi = 0.$$

Boundary conditions

The interface condition (3.2a) is identical to (1.9a). (3.2b) is the dimensionless form of (1.9b). The dimensionless form of the Stefan condition (1.9c) is

$$\vec{e}_y \cdot \vec{n} = \left(\vec{\nabla} T^s - \vec{\nabla} T^l \right) \cdot \vec{n} \tag{A.3}$$

and after inserting the parabolic coordinate expressions for $\vec{\nabla}$, $\vec{e}_y$ and $\vec{n}$ from section 2.1, one finds equation (3.2c). For the mass conservation condition (3.2d), we write (1.13a)

$$\vec{w} \cdot \vec{n} = -\frac{\psi_\xi + \eta'_s \psi_\eta}{\sqrt{(\xi^2 + \eta_s^2)(1 + \eta_s'^2)}}$$

$$= -P_c \vec{e}_y \cdot \vec{n} = -P_c \frac{\xi \eta'_s + \eta_s}{\sqrt{(\xi^2 + \eta_s^2)(1 + \eta_s'^2)}}$$

explicitly in the moving frame of reference and in the rescaled system. The moving frame causes a term containing V on the right hand side, which becomes the P_c-term in its dimensionless form. This leads directly to (3.2d).

The far field boundary conditions (3.3a) and (3.3b) come out naturally when rescaling (1.10a) and (1.10b). For derivation of (3.3c), the definition of $\vec{w}$ and the parabolic coordinate form of $\vec{e}_y$ are inserted into the dimensionless form of (1.14) in the moving frame:

$$\frac{\psi_\eta \vec{e}_\xi - \psi_\xi \vec{e}_\eta}{\sqrt{\xi^2 + \eta^2}} \sim -\frac{\eta \vec{e}_\eta - \xi \vec{e}_\xi}{\sqrt{\xi^2 + \eta^2}} (P_c + P_f) \qquad \eta \to \infty.$$

A.1. Potential flow

Again, a V-term arises on the right hand side due to the moving frame and the V- and U-terms become Péclet number terms after rescaling. From this, the asymptotic behaviour (3.3c) at $\eta \to \infty$ can easily be read.

A.1.2. Asymptotic decomposition

We would like to asymptotically decompose the system (3.9a)-(3.9b) with conditions (3.10a)-(3.10d) and equation (3.1c) with condition (3.13) à la Zauderer [125]. According to section 2.2, we make some definitions:

$$\vec{\vartheta} = \begin{pmatrix} T_\xi \\ T_\eta \end{pmatrix} \qquad \vec{\vartheta}^s = \begin{pmatrix} T_\xi^s \\ T_\eta^s \end{pmatrix} \qquad \vec{v} = \begin{pmatrix} \psi_\xi \\ \psi_\eta \end{pmatrix} \qquad \text{(A.4a)}$$

$$A = \begin{pmatrix} 0 & 1 \\ -1 & 0 \end{pmatrix} \quad B = \begin{pmatrix} -u_0 & -v_0 \\ 0 & 0 \end{pmatrix} \quad C = \begin{pmatrix} F_0 & 0 \\ 0 & 0 \end{pmatrix} \quad \text{(A.4b)}$$

$$u_0 = \psi_\eta + \xi P_f \quad v_0 = -\psi_\xi - P_f(\eta - 1) \quad F_0 = -\mathrm{e}^{-\frac{P_f}{2}(\eta-1)^2}. \quad \text{(A.4c)}$$

Now we have the first order system

$$\vec{\vartheta}_\xi + A\vec{\vartheta}_\eta + B\vec{\vartheta} + C\vec{v} = 0 \qquad \text{in the liquid} \qquad \text{(A.5a)}$$
$$\vec{\vartheta}^s_\xi + A\vec{\vartheta}^s_\eta = 0 \qquad \text{in the solid} \qquad \text{(A.5b)}$$
$$\vec{v}_\xi + A\vec{v}_\eta = 0. \qquad \text{for the stream function} \qquad \text{(A.5c)}$$

The scale transformation $\xi, \eta \to \varepsilon\xi, \varepsilon\eta$ with the small parameter ε is made. We assume εP_f to be of the order of unity, which will later turn out to be an adequate assumption. Apart from that, we expand the variables $\vec{\vartheta}$ and $\vec{v}$ in series of eigenvectors of A:

$$\vec{\vartheta} = M\vec{r}_1 + \varepsilon N\vec{r}_2 \qquad \text{(A.6a)}$$
$$\vec{\vartheta}^s = N^s \vec{r}_2 \qquad \text{(A.6b)}$$
$$\vec{v} = V\vec{r}_1. \qquad \text{(A.6c)}$$

Appendix A. Auxiliary calculations

The eigenvectors $\vec{r}_{1,2}$ can be read from section 2.2. Inserting into the field equations, we get, neglecting terms of the order ε^2

$$M_\xi \vec{r}_1 + \varepsilon N_\xi \vec{r}_2 + i M_\eta \vec{r}_1 - i\varepsilon N_\eta \vec{r}_2$$
$$+ \varepsilon B M \vec{r}_1 + \varepsilon C V \vec{r}_1 = 0 \qquad \text{in the liquid} \qquad \text{(A.7a)}$$
$$N_\xi^s - i N_\eta^s = 0 \qquad \text{in the solid} \qquad \text{(A.7b)}$$
$$V_\xi + i V_\eta = 0. \qquad \text{stream function} \qquad \text{(A.7c)}$$

The intention is to determine the coefficients M, N, N^s and V. The projection operators defined by $P_i \vec{r}_j = \delta_{ij} \vec{r}_i$ can also be found in section 2.2. The reader may verify himself or herself that

$$P_1 B \vec{r}_1 = -a \vec{r}_1 \qquad\qquad P_1 C \vec{r}_1 = \frac{F_0}{2} \vec{r}_1 \qquad \text{(A.8a)}$$
$$P_2 B \vec{r}_1 = a \vec{r}_2 \qquad\qquad P_2 C \vec{r}_1 = -\frac{F_0}{2} \vec{r}_2 \qquad \text{(A.8b)}$$

with $a = \frac{1}{2}(u_0 + i v_0)$ holds. Applying P_1 and P_2 to (A.7a), we find

$$M_\xi + i M_\eta - a M + \frac{F_0}{2} V = 0 \qquad \text{(A.9a)}$$
$$N_\xi - i N_\eta + a M - \frac{F_0}{2} V = 0 \qquad \text{(A.9b)}$$

returning to the original scale. Provided an expression for V, these equations can be solved for M and N subsequently.

To express the boundary conditions (3.10a)-(3.10c) and (3.13) in terms of M, N, N^s and V, we need to calculate their derivative with respect to ξ. At the interface, we have $T = T(\xi, 1 + h)$ and therefore

$$\frac{dT}{d\xi} = T_\xi + h' T_\eta$$

and analog in the solid. From the above definitions we remember

$$T_\xi = -i(M - N)$$

$$T_\eta = M + N$$
$$T_\xi^s = iN^s$$
$$T_\eta^s = N^s$$

and decompose:

$$(1 + ih') M = (1 - ih')(N - N^s) + ih' \quad \text{continuity} \quad \text{(A.10a)}$$
$$(1 - ih') N^s = \frac{i}{2}\sigma\left[\kappa a(\theta)\right]' \quad \text{Gibbs-Thomson} \quad \text{(A.10b)}$$
$$[\xi h]' + (1 - ih')(N - N^s) + (1 + ih') M = 0 \quad \text{Stefan condition.} \quad \text{(A.10c)}$$

Putting (A.10a)-(A.10b) together an, we arrive at

$$M = \frac{i}{2}\frac{\left[(1 + i\xi) h\right]'}{(1 + ih')} \quad \text{(A.11a)}$$

$$(N - N^s) = -\frac{i}{2}\frac{\left[(1 - i\xi) h\right]'}{(1 - ih')} \quad \text{(A.11b)}$$

$$N^s = \frac{i}{2}\frac{\sigma\left[\kappa a(\theta)\right]'}{(1 - ih')} \quad \text{(A.11c)}$$

at the interface. In the interface condition (A.11b), it can be seen that the liquid and the solid domain interact via N as mentioned above. We also see that we will not have to calculate N^s explicitly. It is enough to know its value at the interface, provided a solution in the solid exists. From equation (3.13) we get

$$V = -\frac{iP_f \left[\xi h\right]'}{(1 + ih')} \quad \text{(A.12)}$$

as a boundary condition for equation (A.7c).

We would like to solve the decomposed system (A.9a) and (A.9b) with boundary conditions (A.11a), (A.11b) and (A.11c). But first we need to solve equation (A.7c) with boundary condition (A.12). This

Appendix A. Auxiliary calculations

is done by using the method of characteristics. I. e., we search for a set of coordinates $(s(\xi,\eta), \tau(\xi,\eta))$, in which equation (A.7c) contains derivatives with respect to only one of the new coordinates. We write

$$\frac{\mathrm{d}V}{\mathrm{d}s} = V_\xi \frac{\mathrm{d}\xi}{\mathrm{d}s} + V_\eta \frac{\mathrm{d}\eta}{\mathrm{d}s}$$

and compare coefficients to (A.7c). The result is

$$\frac{\mathrm{d}V}{\mathrm{d}s} = 0 \quad \Rightarrow \quad V = V(\tau)$$

and

$$\frac{\mathrm{d}\xi}{\mathrm{d}s} = 1, \qquad \frac{\mathrm{d}\eta}{\mathrm{d}s} = \mathrm{i}.$$

These equations can be integrated immediately. For comfortability reasons, we wish the interface to be at $s = 0 \Rightarrow \eta(s = 0) = 1$, and accordingly we want $\xi(s = 0) = \tau$. That gives

$$s = -\mathrm{i}\,(\eta - 1) \tag{A.13a}$$
$$\tau = \xi + \mathrm{i}\,(\eta - 1) \tag{A.13b}$$
$$V = -\frac{\mathrm{i}P_f\,[\tau h(\tau)]'}{(1 + \mathrm{i}h')}. \tag{A.14}$$

Equation (A.9a) has the same characteristics and it reads

$$M_s - \frac{P_f}{2}(2s + \tau)\,M + \frac{\mathrm{i}P_f\,[\tau h(\tau)]'}{2(1 + \mathrm{i}h')} e^{\frac{P_f}{2}s^2} = 0. \tag{A.15}$$

The solution to this equation is given in (3.17a), and it can easily be found by first ignoring the inhomogeneity, and then adding a particular solution found by variation of constants. The constant of integration is deduced from condition (A.11a). In equation (A.9b) we use the characteristics

$$\bar{s} = \mathrm{i}\,(\eta - 1) \tag{A.16a}$$

A.1. Potential flow

$$\bar{\tau} = \xi - \mathrm{i}\,(\eta - 1) \tag{A.16b}$$

with the interface at $\bar{s} = 0$ and $\bar{\tau} = \xi\,(\bar{s} = 0)$. In these coordinates equation (A.9b) reads

$$N_{\bar{s}} = -\frac{P_f}{2}\bar{\tau}M(-\bar{s}, \bar{\tau} + 2\bar{s}) + \frac{\mathrm{i}P_f\left[(\bar{\tau} + 2\bar{s})h(\bar{\tau} + 2\bar{s})\right]'}{2\,(1 + \mathrm{i}h'(\bar{\tau} + 2\bar{s}))}\mathrm{e}^{\frac{P_f}{2}\bar{s}^2}, \tag{A.17}$$

which can be integrated immediately. The boundary condition for N is derived by putting together (A.11b) and (A.11c):

$$N(\bar{s} = 0) = \frac{\mathrm{i}}{2}\frac{\sigma\kappa'(\bar{\tau})}{(1 - \mathrm{i}h'(\bar{\tau}))} - \frac{\mathrm{i}}{2}\frac{\left[(1 - \mathrm{i}\bar{\tau})\,h(\bar{\tau})\right]'}{(1 - \mathrm{i}h'(\bar{\tau}))}. \tag{A.18}$$

This leads to the solution given in (3.17b).

A.1.3. WKB solution

Derivation of equations (3.21) and (3.22)
We start from (3.20) with the prime denoting derivatives with respect to ξ or ξ', depending on the concerning term being outside or inside an integral respectively. Writing M in these terms, we get

$$M\left(\frac{1}{2}(\xi - \xi'), \xi'\right) = \frac{\mathrm{i}}{2z(\xi')}$$
$$\times \mathrm{e}^{\frac{P_f}{8}(\xi^2 - \xi'^2)}\left[\left[(1 + \mathrm{i}\xi')\,h\right]' - \frac{2\,[\xi'h]'}{\xi'}\left(1 - \mathrm{e}^{-\frac{P_f}{4}(\xi - \xi')\xi'}\right)\right]$$

and

$$\frac{\partial}{\partial \xi}M\left(\frac{1}{2}(\xi - \xi'), \xi'\right) = \frac{P_f\xi}{4}M\left(\frac{1}{2}(\xi - \xi'), \xi'\right) - \frac{\mathrm{i}P_f}{4}\frac{[\xi'h]'}{z(\xi')}\mathrm{e}^{\frac{P_f}{8}(\xi - \xi')^2}.$$

Putting this into (3.20), it turns out that

$$\tilde{F}'(\xi) = -2\mathrm{i}\bar{z}(\xi)\int_\xi^{\mathrm{i}\infty}\frac{\partial}{\partial \xi}M\left(\frac{1}{2}(\xi - \xi'), \xi'\right)\mathrm{d}\xi'$$

Appendix A. Auxiliary calculations

$$= -2i\bar{z}(\xi)\left[\frac{\partial}{\partial \xi}\int_\xi^{i\infty} M\left(\frac{1}{2}(\xi-\xi'),\xi'\right)d\xi' + M(0,\xi)\right]$$

$$= -2i\bar{z}(\xi)\frac{\partial}{\partial \xi}\int_\xi^{i\infty} M\left(\frac{1}{2}(\xi-\xi'),\xi'\right)d\xi' + \frac{\bar{z}(\xi)}{z(\xi)}\left[(1+i\xi)h(\xi)\right]'.$$

This can be integrated by parts. Using

$$\left[\frac{\bar{z}}{z}\right]' = -2i\frac{h''}{z^2}$$

we arrive at (3.21). We use (3.21) to eliminate the first term on the right hand side of (3.20):

$$\tilde{F}'(\xi) = \frac{P_f}{4}\xi\tilde{F}(\xi) - \frac{P_f}{4}\xi\frac{\bar{z}(\xi)}{z(\xi)}(1+i\xi)h(\xi) - \frac{i}{2}P_f\xi\int_\xi^\xi \frac{h''(\xi')}{z^2(\xi')}(1+i\xi')h(\xi')d\xi'$$
$$- \frac{P_f}{2}\xi\int_\xi^\xi h''(\xi')\int_{\xi'}^{i\infty} M\left(\frac{1}{2}(\xi'-\xi''),\xi''\right)d\xi''\,d\xi' - \frac{P_f}{2}\bar{z}(\xi)\int_\xi^{i\infty}\frac{[\xi'h]'}{z(\xi')}e^{\frac{P_f}{8}(\xi-\xi')^2}d\xi'.$$

This is altered again, using the three identities

$$\int_\xi^{i\infty}\frac{[\xi'h]'}{z(\xi')}e^{\frac{P_f}{8}(\xi-\xi')^2}d\xi' \stackrel{\text{P. I.}}{=} \frac{\xi h(\xi)}{z(\xi)}$$
$$-\int_\xi^{i\infty}\left[\xi'h(\xi')\left(\frac{(\xi'-\xi)P_f}{4z(\xi')} - \frac{ih''(\xi')}{z^2(\xi')}\right)e^{\frac{P_f}{8}(\xi-\xi')^2}\right]d\xi'$$

$$\tilde{F}'(\xi) - \frac{P_f}{4}\xi\tilde{F}(\xi) = e^{\frac{P_f}{8}\xi^2}\frac{d}{d\xi}\left[\tilde{F}(\xi)e^{-\frac{P_f}{8}\xi^2}\right]$$

$$2h(\xi) - (1+i\xi)h(\xi) = (1-i\xi)h(\xi)$$

A.1. Potential flow

and we get

$$e^{\frac{P_f}{8}\xi^2}\frac{d}{d\xi}\left[\tilde{F}(\xi)e^{-\frac{P_f}{8}\xi^2}\right] = \frac{P_f}{4}\frac{\bar{z}(\xi)}{z(\xi)}\xi\left(1-i\xi\right)h(\xi)$$

$$+\frac{P_f}{2}\bar{z}(\xi)\int\limits_{\xi}^{i\infty}\left[\xi'h(\xi')\left(\frac{(\xi'-\xi)P_f}{4z(\xi')}-\frac{ih''(\xi')}{z^2(\xi')}\right)e^{\frac{P_f}{8}(\xi-\xi')^2}\right]d\xi'$$

$$-\frac{i}{2}P_f\xi\int\limits^{\xi}\frac{h''(\xi')}{z^2(\xi')}(1+i\xi')h(\xi')d\xi'$$

$$-\frac{P_f}{2}\xi\int\limits^{\xi}h''(\xi')\int\limits_{\xi'}^{i\infty}M\left(\frac{1}{2}(\xi'-\xi''),\xi''\right)d\xi''\,d\xi'\,.$$

Integrating this, we arrive at the final equation (3.22).

Finding the WKB solution (3.24)
In the WKB ansatz

$$\begin{aligned}h(\xi) &= \exp\left(\frac{1}{\varepsilon}\sum_{k=0}^{\infty}\varepsilon^k S_k(\xi)\right) \\ &\approx \exp\left(\frac{1}{\varepsilon}S_0(\xi)+S_1(\xi)+\frac{1}{\varepsilon}S_0'(\xi)(\xi'-\xi)\right),\end{aligned} \quad (A.19)$$

the small parameter must be $\varepsilon = \sqrt{\sigma}$ for consistency. Using (A.19), we can write the full WKB equation, sorted by powers of ε. The following simplification is used in (3.56):

$$\frac{P_f}{4}e^{\frac{P_f}{8}\xi^2}\int^{\xi}e^{-\frac{P_f}{8}\xi'^2+S_1(\xi')}\xi'(1-i\xi')e^{\frac{1}{\varepsilon}S_0(\xi')}d\xi'$$

$$\propto \frac{P_f}{4}(1-i\xi)\,\xi e^{\frac{1}{\varepsilon}S_0(\xi)+S_1(\xi)}\frac{\varepsilon}{S_0'(\xi)}. \quad (A.20)$$

Since we took into account only S_0 and S_1 in the ansatz, we neglect terms of the order ε^2:

Appendix A. Auxiliary calculations

$$0 = S_0'^2 + \sqrt{1+\xi^2}\,(1-i\xi)$$
$$+ \varepsilon\left(\frac{\xi}{1+\xi^2}S_0' + 2S_0'S_1' + S_0'' + \frac{P_f}{4S_0'}\sqrt{1+\xi^2}\,(1-i\xi)\,\xi\right). \quad (A.21)$$

From the zero order we get

$$\varepsilon^0: \quad S_0'^2 = -\sqrt{1+\xi^2}\,(1-i\xi)$$

and

$$S_0(\xi) = i\int_{-i}^{\xi} (1+i\xi')^{\frac{1}{4}}(1-i\xi')^{\frac{3}{4}}\,d\xi'.$$

The first order equation reads

$$\frac{\xi}{1+\xi^2}S_0' + 2S_0'S_1' + S_0'' - \frac{P_f}{4}\xi S_0' = 0.$$

From this, S_1 is easily calculated:

$$\varepsilon^1: \quad S_1'(\xi) = \frac{P_f}{8}\xi - \frac{1}{2}\left[\frac{S_0''}{S_0'} + \frac{\xi}{1+\xi^2}\right]$$

$$S_1(\xi) = \frac{P_f}{8}\int_{-i}^{\xi}\xi'\,d\xi' - \frac{1}{2}\int^{\xi}\left[\frac{S_0''}{S_0'} + \frac{\xi'}{1+\xi'^2}\right]d\xi'$$

$$= \frac{P_f}{16}\xi^2 + \frac{P_f}{16} - \frac{1}{2}\ln S_0' - \frac{1}{4}\ln(1+\xi^2)$$

$$= \frac{P_f}{16}\xi^2 + \frac{P_f}{16} + \ln\left[(1+i\xi)^{-\frac{3}{8}}(1-i\xi)^{-\frac{5}{8}}\right] + \text{const.}$$

Inserting this into the ansatz (A.19) for h, we get the desired asymptotic approximation (3.24) far from the singularity.

A.1.4. Local equation

Stretching transformation
The formulas

$$h' = -i\dot{\phi} \tag{A.22a}$$
$$h'' = -\sigma^{-\alpha}\ddot{\phi} \tag{A.22b}$$
$$(1 - i\xi)\, h = \sigma^{2\alpha}\phi t \tag{A.22c}$$

are easily derived from (3.27a)-(3.27b). At first, the fourfold surface tension anisotropy function $a(\theta)$ from (2.10b) is transformed. For this purpose, the following expressions are needed:

$$[\xi - (1+h)h']^2 = -\left[1 - \dot{\phi} + \sigma^\alpha(t - \phi\ddot{\phi})\right]^2 \approx -(1 - \dot{\phi})^2$$
$$[1 + [\xi h]']^2 = \left[1 - \dot{\phi} + \sigma^\alpha [t\phi]'\right]^2 \approx (1 - \dot{\phi})^2$$
$$[\xi^2 + (1+h)^2]^2 = \left[2\sigma^\alpha(t + \phi) + \sigma^{2\alpha}(\phi^2 - t^2)\right]^2 \approx 4\sigma^{2\alpha}(t + \phi)^2$$
$$(1 + h'^2)^2 = (1 - \dot{\phi}^2)^2.$$

Only the leading order of each expression is inserted into $a(\theta)$. Consequently, in the numerator there is only one term, and this term is $\mathcal{O}(1)$. The term $\xi^2 + (1+h)^2$ from the third line of the above equations is responsible for the divergence of $a(\theta)$, because its leading order is $\mathcal{O}(\sigma^\alpha)$:

$$a(\theta) = 1 - \beta\left[\mathcal{X} + 8\underbrace{\frac{(1 - \dot{\phi})^4}{4\sigma^{2\alpha}(t + \phi)^2(1 - \dot{\phi}^2)^2}}_{\gg 1}\right] \approx 1 - \frac{2\beta\sigma^{-2\alpha}(1 - \dot{\phi})^2}{(t + \phi)^2(1 + \dot{\phi})^2}. \tag{A.23}$$

Since $b = \beta\sigma^{-2\alpha}$ is assumed to be $\mathcal{O}(1)$, the remaining 1 cannot be neglected.

Appendix A. Auxiliary calculations

Inserting (A.22a)-(A.22c) into the left hand side of (3.26), using the curvature (2.7) and keeping only terms of the lowest order of σ we find

$$\kappa = \frac{\sigma^{-\frac{3}{2}\alpha}}{\sqrt{2t+2\phi}} \left(\frac{\ddot{\phi}}{\left(1-\dot{\phi}^2\right)^{\frac{3}{2}}} + \frac{1+\dot{\phi}}{(2t+2\phi)\sqrt{1-\dot{\phi}^2}} \right) \qquad (A.24)$$

$$\tilde{F} = \sigma^{2\alpha} \left[\frac{1}{\sqrt{2t+2\phi}} \left(\frac{\ddot{\phi}}{\left(1-\dot{\phi}^2\right)^{\frac{3}{2}}} + \frac{1+\dot{\phi}}{(2t+2\phi)\sqrt{1-\dot{\phi}^2}} \right) - \phi t \right] a(\theta) \qquad (A.25)$$

for the curvature. Transforming the right hand side of (3.26) takes a bit more effort. Calculating the terms one by one, we note some identities:

$$m_1 := \xi'^2 - \xi''^2 = 2\sigma^\alpha (t' - t'') - \sigma^{2\alpha} \left(t'^2 - t''^2\right)$$
$$m_2 := \xi'' (\xi'' - \xi') = \sigma^\alpha (t'' - t') + \sigma^{2\alpha} \left(t't'' - t''^2\right)$$
$$m_3 := (\xi' - \xi'')^2 = -\sigma^{2\alpha} (t' - t'')^2 \stackrel{!}{=} m_1 + 2m_2$$
$$[(1+i\xi'')\,h]' = i\left(\sigma^\alpha \phi - (2 - \sigma^\alpha t'')\dot{\phi}\right)$$
$$[\xi''h]' = \sigma^\alpha \phi - (1 - \sigma^\alpha t'')\dot{\phi}$$
$$z = 1 + \dot{\phi}$$
$$\bar{z} = 1 - \dot{\phi}.$$

With these formulas, the integrand can be calculated quite comfortably. Remembering $d\xi = i\sigma^\alpha dt$, we arrive at

$$\tilde{F} = \frac{P_f}{4}\sigma^{3\alpha} \int_\infty^t \int^{t'} \frac{1-\dot{\phi}(t')}{1+\dot{\phi}(t'')} \left[e^{\frac{P_f}{4}\sigma^\alpha (t'-t'')} \left(t''\dot{\phi} - \phi \right) \right. \qquad (A.26)$$
$$\left. + 2\dot{\phi} \times (t' - t'') \right] dt'' dt'.$$

A.1. Potential flow

Here, the functions with σ^α in the denominator have been expanded in a *Taylor series* up to the first order. The exponential function containing m_3 is approximately 1. The lowest order of σ occuring is 3α (before division by $\sigma^{2\alpha}$, which has to be done to reach the final equation). Now, we have all the ingredients to write the full nonlinear eigenvalue problem (3.30) determining σ.

Matching to the WKB solution (3.24)

The dominant balance in (3.30) is

$$\phi t + P_1 \int^t \frac{1-\dot{\phi}(t')}{1+\dot{\phi}(t')}\phi(t')t'\mathrm{d}t' \sim \frac{1}{(2t)^{3/2}}. \quad (A.27)$$

This does not go to the real "beyond-all-orders regime", and an asymptotic analysis solution to this equation is never going to match properly, no matter how many orders are calculated. Instead, we approximately calculate the terms of the order $\sigma^{4\alpha}$ in the local equation, which were neglected in (3.30). These are

$$-\frac{P_f}{4}\int^t \left(1-\dot{\phi}\right)\int_{t'}^\infty \frac{1}{1+\dot{\phi}(t'')}\left(-\dot{\phi}(t'')t' - \ddot{\phi}(t'')t't''\right)\mathrm{d}t''\mathrm{d}t'$$

$$=\frac{P_f}{4}\int^t \left(1-\dot{\phi}\right)\int_{t'}^\infty \frac{1}{1+\dot{\phi}(t'')}t'\frac{\mathrm{d}}{\mathrm{d}t''}\left[\dot{\phi}(t'')t''\right]\mathrm{d}t''\mathrm{d}t'$$

$$=-\frac{P_f}{4}\int^t \frac{1-\dot{\phi}(t')}{1+\dot{\phi}(t')}\dot{\phi}(t')t'^2\mathrm{d}t'.$$

Using $t \gg \phi$, we find

$$\frac{\ddot{\phi}}{\sqrt{2t}} + \frac{\dot{\phi}}{(2t)^{\frac{3}{2}}} = \phi t + \int^t \frac{1-\dot{\phi}(t')}{1+\dot{\phi}(t')}\dot{\phi}(t')\left(P_1 t' - P_2 t'^2\right)\mathrm{d}t'.$$

Appendix A. Auxiliary calculations

Now we approximate

$$\frac{1-\dot\phi(t')}{1+\dot\phi(t')} \approx 1 \qquad t \to \infty$$

and differentiate once with respect to t:

$$\frac{d^3\phi}{dt^3} - \sqrt{2}t^{\frac{3}{2}}\dot\phi - \sqrt{2t}\phi\left(1 + P_1 t - P_2 t^2\right) = 0. \qquad (A.28)$$

We make the common asymptotic analysis ansatz $\phi(t) = e^{S(t)}$ and find the dominant balance

$$S'^3 \sim \sqrt{2}t^{\frac{3}{2}}S'$$

yielding

$$S = -2^{\frac{1}{4}}\frac{4}{7}t^{\frac{7}{4}} + c.$$

Reinserting, we find the dominant behaviour of the next order.

$$2c'\sqrt{2}t^{\frac{3}{2}} \sim \sqrt{2t}\left(1 + P_1 t - P_2 t^2\right) - \frac{9}{4}\sqrt{2t}$$

$$c' \sim -\frac{5}{8t} + \frac{P_1}{2} - \frac{P_2}{2}t$$

$$c = \ln t^{-\frac{5}{8}} + \frac{P_1}{2}t - \frac{P_2}{4}t^2.$$

Putting $S + c$ into the ansatz, we find (3.33).

We proceed, putting the stretching transformation into the prefactor of the exponential function in the WKB solution (3.24), keeping only leading orders:

$$(1 + i\xi)^{-\frac{3}{8}}(1 - i\xi)^{-\frac{5}{8}} \approx \sigma^{-\frac{5}{28}} 2^{-\frac{3}{8}} t^{-\frac{5}{8}}.$$

Hence, the prefactor is reproduced. Now we calculate S_0:

$$S_0 = -\int_0^t \sigma^{\frac{3}{14}}t'^{\frac{3}{4}} 2^{\frac{1}{4}} \sigma^{\frac{2}{7}} dt' = -2^{\frac{1}{4}}\frac{4}{7}t^{\frac{7}{4}}\sqrt{\sigma}.$$

Hence, the first term of the exponent matches. The other terms of the exponent are also matching each other:

$$\frac{P_f}{16}\left(\xi^2+1\right)=\frac{P_f}{16}\left(2\sigma^\alpha t-\sigma^{2\alpha}t^2\right)=\frac{P_1}{2}t-\frac{P_2}{2}t^2.$$

Considering $h(\xi)=\sigma^\alpha \phi(t)$, the constants must be related by

$$A_1=B_1 2^{-\frac{3}{8}}\sigma^{-\frac{13}{28}}. \qquad (A.29)$$

To evaluate the physical condition of a smooth tip, we need

$$\left.\frac{d\eta_s}{d\xi}\right|_{\xi=0}=\left.\frac{dh}{d\xi}\right|_{\xi=0}$$

$$=B_1 e^{\frac{S_0(0)}{\sqrt{\sigma}}+\frac{P_f}{16}}\left[\overbrace{\frac{S_0'(0)}{\sqrt{\sigma}}}^{=i}+\frac{i}{4}\right]$$

$$=iA_1 2^{\frac{3}{8}}\sigma^{\frac{13}{28}} e^{\frac{S_0(0)}{\sqrt{\sigma}}+\frac{P_f}{16}}\left[\frac{1}{\sqrt{\sigma}}+\frac{1}{4}\right]$$

and the condition reads

$$\mathfrak{Re}\left[\left.\frac{d\eta_s}{d\xi}\right|_{\xi=0}\right]=\mathfrak{Im}\left(A_1\right)2^{\frac{3}{8}}\sigma^{\frac{13}{28}}e^{\frac{-4\sqrt[4]{2}}{7\sqrt{\sigma}}+\frac{P_f}{16}}\left[\frac{1}{\sqrt{\sigma}}+\frac{1}{4}\right]\stackrel{!}{=}0 \qquad (A.30)$$

leading to the selection criterion (3.34).

A.2. Oseen flow

A.2.1. Ivantsov solution

Stream function
Since (3.35) is linear, we will construct a solution by linear combination of two contributions. For the first part $\vec{w}^1$ we make the ansatz

$$\vec{w}^1=Pr\vec{\nabla}\chi-\vec{v}_1\chi \qquad (A.31)$$

Appendix A. Auxiliary calculations

with $\vec{v}_1 = -(P_c + P_f)\vec{e}_y = -P_4\vec{e}_y$ and a dependent function χ. We insert it into (3.35) without the pressure term:

$$\underbrace{\left(\vec{v}_1 \cdot \vec{\nabla}\right) Pr\vec{\nabla}\chi}_{=Pr\vec{\nabla}(\vec{v}_1 \cdot \vec{\nabla}\chi)} - \underbrace{\left(\vec{v}_1 \cdot \vec{\nabla}\right) \vec{v}_1\chi}_{=\vec{v}_1(\vec{v}_1 \cdot \vec{\nabla}\chi)} = \underbrace{Pr\Delta\left(Pr\vec{\nabla}\chi\right)}_{=Pr\vec{\nabla}(Pr\Delta\chi)} - \underbrace{Pr\Delta\left(\vec{v}_1\chi\right)}_{=\vec{v}_1(Pr\Delta\chi)}. \tag{A.32}$$

We get

$$Pr\vec{\nabla}\left(Pr\Delta\chi - \vec{v}_1 \cdot \vec{\nabla}\chi\right) = \vec{v}_1\left(Pr\Delta\chi - \vec{v}_1 \cdot \vec{\nabla}\chi\right) \tag{A.33}$$

and notice that this will only hold if

$$Pr\Delta\chi + P_4\vec{e}_y \cdot \vec{\nabla}\chi = 0. \tag{A.34}$$

To obtain a general solution to equation (3.35), we need to add another term $\vec{w}^2$ and we make a potential flow ansatz:

$$\vec{w}^2 = -\vec{\nabla}\varphi. \tag{A.35}$$

One requirement to $\vec{w}^2$ is to be linearly independent of $\vec{w}^1$. Let's see if our ansatz can succeed,

$$0 = \left(\vec{v}_1 \cdot \vec{\nabla}\right)\vec{\nabla}\varphi - \vec{\nabla}p - Pr\Delta\left(\vec{\nabla}\varphi\right) = \vec{\nabla}\left(\vec{v}_1 \cdot \vec{\nabla}\varphi - p - Pr\Delta\varphi\right).$$

If we assume a *Laplace equation* to be valid for φ, we may write a full set of equations determining the solution $\vec{w}$

$$0 = Pr\Delta\chi + P_4\vec{e}_y \cdot \vec{\nabla}\chi \tag{A.36a}$$
$$0 = p + P_4\vec{e}_y \cdot \vec{\nabla}\varphi \tag{A.36b}$$
$$0 = \Delta\varphi \tag{A.36c}$$

and

$$\vec{w} = \vec{w}^1 + \vec{w}^2 = Pr\vec{\nabla}\chi - \vec{v}_1\chi - \vec{\nabla}\varphi. \tag{A.37}$$

A.2. Oseen flow

As zeroth approximation, we assume a perfectly parabolic interface and look for appropriate solutions [66]. In parabolic coordinates, the interface is at $\eta_s = 1$. Ananth and Gill [10] noted, that for a parabolic interface, similarity solutions exist. Therefore, we want solutions of the kind $\chi = \chi(\eta)$ and $\varphi = \varphi(\eta)$. Using this, equations (A.36a) and (A.36c) become quite simple:

$$0 = Pr \chi_{\eta\eta} + P_4 \eta \chi_\eta, \qquad (A.38)$$
$$0 = \varphi_{\eta\eta}. \qquad (A.39)$$

These equations are integrated right away. We use the abbreviations (3.39),

$$\chi(\eta) = c_3 \int_1^\eta e^{-\tilde{\alpha}^2 \eta'^2} d\eta' + c_4, \qquad (A.40)$$

$$\varphi(\eta) = c_5 \eta + c_6. \qquad (A.41)$$

Writing the components of $\vec{w}$, we get expressions for the derivatives of the stream function:

$$\psi_\eta = -P_4 \chi \xi,$$
$$\psi_\xi = c_5 - Pr\, c_3 e^{-\tilde{\alpha}^2 \eta^2} - P_4 \chi \eta.$$

We may now apply the boundary conditions (3.2d), (3.3c) and (3.38) consecutively to determine the corresponding values of the integration constants:

$$c_4 = -\frac{P_c}{P_4} \qquad \text{from (3.38)}$$

$$c_3 = -\frac{P_f}{P_4} \frac{2\tilde{\alpha}}{\sqrt{\pi}\, \text{erfc}(\tilde{\alpha})} \qquad \text{from (3.3c)}$$

$$c_5 = -\frac{P_f e^{-\tilde{\alpha}^2}}{\tilde{\alpha}\sqrt{\pi}\, \text{erfc}(\tilde{\alpha})} \qquad \text{from (3.2d)}.$$

Appendix A. Auxiliary calculations

This can but need not be the order of progression. c_6 does not appear explicitly in the solution. It is just an arbitrary base value for the potential function $\varphi(\eta)$ and can be set to zero:

$$c_6 = 0\,.$$

We now have all the ingredients to write the solution (3.40b), (3.40a) in a convenient form. Note, that the η-component of $\vec{w}$ can be expressed using the ξ-component:

$$\psi_\xi(\xi,\eta) = \frac{\eta}{\xi}\psi_\eta(\eta) + \frac{P_f \tilde{a}}{2\tilde{a}^2}\left[e^{\tilde{a}^2(1-\eta^2)} - 1\right]. \tag{A.42}$$

Temperature field

The derivation of (3.41) is analogue to the potential flow case. The remaining task is to calculate $I_1(\eta)$. We use

$$\frac{\mathrm{d}}{\mathrm{d}\omega}\operatorname{erfc}(\tilde{a}\omega) = -\frac{2\tilde{a}}{\sqrt{\pi}}e^{-\tilde{a}^2\omega^2}$$

and integration by parts several times. Regarding (3.40b), on term to integrate is

$$t_1 = \int_1^\eta \omega \left[P_c + P_f\left(1 - \frac{\operatorname{erfc}(\tilde{a}\omega)}{\operatorname{erfc}(\tilde{a})}\right)\right]\mathrm{d}\omega$$

$$\stackrel{\text{p. I.}}{=} \frac{\eta^2}{2}\left[P_c + P_f\left(1 - \frac{\operatorname{erfc}(\tilde{a}\eta)}{\operatorname{erfc}(\tilde{a})}\right)\right] - \frac{P_c}{2} + \frac{1}{2}P_f \int_1^\eta \omega^2 \frac{\mathrm{d}}{\mathrm{d}\omega}\frac{\operatorname{erfc}(\tilde{a}\omega)}{\operatorname{erfc}(\tilde{a})}\mathrm{d}\omega$$

$$= \frac{\eta^2}{2}\left[P_c + P_f\left(1 - \frac{\operatorname{erfc}(\tilde{a}\eta)}{\operatorname{erfc}(\tilde{a})}\right)\right] - \frac{P_c}{2} - \frac{1}{2}P_f \tilde{a}\underbrace{\int_1^\eta \omega^2 e^{\tilde{a}^2(1-\omega^2)}\mathrm{d}\omega}_{=t_2}\,.$$

We continue with the next term to integrate. We use the formula

$$\tilde{a}\int_1^\eta e^{\tilde{\alpha}^2(1-\omega^2)}d\omega = 1 - \frac{\text{erfc}(\tilde{\alpha}\eta)}{\text{erfc}(\tilde{\alpha})}.$$

It can be verified by substitution:

$$t_2 = -\frac{1}{2}P_f\tilde{a}\int_1^\eta \omega^2 e^{\tilde{\alpha}^2(1-\omega^2)}d\omega$$

$$\overset{\text{p. I.}}{=} -\frac{1}{2}P_f\tilde{a}\left[-\frac{\omega}{2\tilde{\alpha}^2}e^{\tilde{\alpha}^2(1-\omega^2)}\bigg|_1^\eta + \frac{1}{2\tilde{\alpha}^2}\int_1^\eta e^{\tilde{\alpha}^2(1-\omega^2)}d\omega\right]$$

$$= \frac{P_f\tilde{a}}{4\tilde{\alpha}^2}\left[\eta e^{\tilde{\alpha}^2(1-\eta^2)} - 1\right] - \frac{P_f}{4\tilde{\alpha}^2}\left[1 - \frac{\text{erfc}(\tilde{\alpha}\eta)}{\text{erfc}(\tilde{\alpha})}\right].$$

Now, only one term remains:

$$t_3 = \frac{P_f\tilde{a}}{2\tilde{\alpha}^2}\int_1^\eta \left(e^{\tilde{\alpha}^2(1-\omega^2)} - 1\right)d\omega$$

$$= -\frac{P_f\tilde{a}}{2\tilde{\alpha}^2}(\eta - 1) + \frac{P_f}{2\tilde{\alpha}^2}\left[1 - \frac{\text{erfc}(\tilde{\alpha}\eta)}{\text{erfc}(\tilde{\alpha})}\right].$$

We get (3.42) with $I_1(\eta) = t_1 + t_3$.

A.2.2. Asymptotic decomposition

Power series solution of (3.46b)
We look for an analytical solution to equation (3.46b). Let us try a separation ansatz for the function χ:

$$\chi(\xi, \eta) = \Xi(\xi)E(\eta). \qquad (A.43)$$

Appendix A. Auxiliary calculations

This is inserted into equation (3.46b):

$$\frac{\Xi''}{\Xi} - 2\tilde{\alpha}^2 \xi \frac{\Xi'}{\Xi} = -\frac{E''}{E} - 2\tilde{\alpha}^2 \eta \frac{E'}{E} \stackrel{!}{=} c_7.$$

Both sides have to be constant, because each side depends only on one of the independent variables:

$$\Xi'' - 2\tilde{\alpha}^2 \xi \Xi' - c_7 \Xi = 0.$$

This equation is solved in general by a power series in ξ:

$$\Xi(\xi) = \sum_{k=0}^{\infty} b_k \, \xi^k$$

$$\Xi'(\xi) = \sum_{k=0}^{\infty} k \, b_k \, \xi^{k-1} = \sum_{k=1}^{\infty} k \, b_k \, \xi^{k-1} = \sum_{k=0}^{\infty} (k+1) \, b_{k+1} \, \xi^k$$

$$\Xi''(\xi) = \sum_{k=0}^{\infty} (k+1) \, k \, b_{k+1} \, \xi^{k-1} = \sum_{k=1}^{\infty} (k+1) \, k \, b_{k+1} \, \xi^{k-1}$$

$$= \sum_{k=0}^{\infty} (k+1)(k+2) \, b_{k+2} \, \xi^k.$$

We insert this into the differential equation and collect the coefficients of like powers of ξ:

$$0 = \sum_{k=0}^{\infty} \underbrace{\left((k+1)(k+2) \, b_{k+2} - k \, b_k \, 2\tilde{\alpha}^2 - c_7 \, b_k \right)}_{\stackrel{!}{=} 0} \xi^k.$$

The coefficients of like powers must vanish identically. The result is a recursive formula determining the coefficients b_k:

$$b_{k+2} = \frac{2\tilde{\alpha}^2 k + c_7}{(k+1)(k+2)} b_k. \qquad \text{(A.44a)}$$

The coefficients b_0 and b_1 must be gained from boundary conditions. They are identified with the integration constants of the differential equation. Subsequently, the remaining b_k with odd and even k can be calculated one by one using (A.44a). In the same manner one can derive a solution

$$E(\eta) = \sum_{l=0}^{\infty} e_l \eta^l$$

with a similar recursion relation:

$$e_{l+2} = -\frac{2\tilde{\alpha}^2 l - c_7}{(l+1)(l+2)} e_l. \tag{A.44b}$$

Indeed, we found the exact solution (3.48) for χ.

Calculation of an asymptotically consistent correction to the flow velocity field

The system (3.46a)-(3.46b) is equivalent to four first order equations. We will treat it as two times two first order equations and decompose two first order 2×2-matrix equations subsequently instead of decomposing one first order 4×4-matrix equation. Our choice of procedure is clearer and the results are the same. Equation (3.46b) is written as a first order system:

$$\vec{\mu}_\xi + A\vec{\mu}_\eta + 2\tilde{\alpha}^2 F(\xi,\eta)\vec{\mu} = 0. \tag{A.45}$$

with

$$\vec{\mu} = \begin{pmatrix} \chi_\xi \\ \chi_\eta \end{pmatrix}, \qquad A = \begin{pmatrix} 0 & 1 \\ -1 & 0 \end{pmatrix}, \qquad F(\xi,\eta) = \begin{pmatrix} -\xi & \eta \\ 0 & 0 \end{pmatrix}.$$

We repeatedly refer to section 2.2. $\vec{\mu}$ is expanded appropriately in eigenvectors $\vec{r}_1$ and $\vec{r}_2$ of A.

$$\vec{\mu} = \lambda^{(1)} \vec{r}_1 + \varepsilon \lambda^{(2)} \vec{r}_2 \tag{A.46}$$

ε is a small parameter, which is used for the scale transformation $(\xi,\eta) \to (\varepsilon\xi,\varepsilon\eta)$:

$$\left(\lambda^{(1)}_\xi + i\lambda^{(1)}_\eta\right) \vec{r}_1 + \varepsilon \left(\lambda^{(2)}_\xi - i\lambda^{(2)}_\eta\right) \vec{r}_2 + 2\varepsilon\tilde{\alpha}^2 F(\varepsilon\xi,\varepsilon\eta)\lambda^{(1)}\vec{r}_1 = 0.$$
$$\tag{A.47}$$

Appendix A. Auxiliary calculations

This is a new form of (A.45), where we neglected $\mathcal{O}(\varepsilon^2)$. The term $2\varepsilon\tilde{\alpha}^2\, F\left(\varepsilon\xi,\varepsilon\eta\right)\lambda^{(1)}\vec{r}_1$ is the smallest term kept. This equation is projected onto the invariant subspaces of A:

$$\lambda^{(1)}_\xi + \mathrm{i}\lambda^{(1)}_\eta = \tilde{\alpha}^2\left(\xi - \mathrm{i}\eta\right)\lambda^{(1)}, \tag{A.48a}$$

$$\lambda^{(2)}_\xi - \mathrm{i}\lambda^{(2)}_\eta = -\tilde{\alpha}^2\left(\xi - \mathrm{i}\eta\right)\lambda^{(1)}. \tag{A.48b}$$

We already returned to the original scale. These equations are a coupled system of first order partial differential equations. It can be solved by means of the *method of characteristics*. The characteristic coordinates are given in (3.15a) and (3.15b) for (A.48a) and (A.48b) respectively. The idea is to integrate on characteristic lines in the $\mathbb{C}^2$ along which the partial differential equations become ordinary,

$$\lambda^{(1)}_s = \tilde{\alpha}^2\left[2s - \mathrm{i}\left(1 + \mathrm{i}\tau\right)\right]\lambda^{(1)}, \tag{A.49a}$$

$$\lambda^{(2)}_{\bar{s}} = \tilde{\alpha}^2\mathrm{i}\left(1 + \mathrm{i}\bar{\tau}\right)\lambda^{(1)}. \tag{A.49b}$$

These equations can be integrated immediately,

$$\lambda^{(1)}(s,\tau) = c_8(\tau)\mathrm{e}^{\tilde{\alpha}^2\left[s^2 - \mathrm{i}s(1+\mathrm{i}\tau)\right]}, \tag{A.50a}$$

$$\lambda^{(2)}(\bar{s},\bar{\tau}) = \tilde{\alpha}^2\mathrm{i}\left(1 + \mathrm{i}\bar{\tau}\right)\int_0^{\bar{s}} \lambda^{(1)}(-\omega, \bar{\tau} + 2\omega)\,\mathrm{d}\omega + c_9(\bar{\tau})$$

$$= \tilde{\alpha}^2\mathrm{i}\left(1 + \mathrm{i}\bar{\tau}\right)\int_0^{\bar{s}} c_8(\bar{\tau} + 2\omega)\mathrm{e}^{\tilde{\alpha}^2\left[-\omega^2 + \mathrm{i}\omega(1+\mathrm{i}\bar{\tau})\right]}\,\mathrm{d}\omega + c_9(\bar{\tau}). \tag{A.50b}$$

Now, we can write a proper form of (3.46a):

$$\psi_{\xi\xi} + \psi_{\eta\eta} = -P_f\left[\xi\left(\lambda^{(1)} + \lambda^{(2)}\right) - \mathrm{i}\eta\left(\lambda^{(1)} - \lambda^{(2)}\right)\right]. \tag{A.51}$$

We use

$$\vec{v} = \begin{pmatrix} \psi_\xi \\ \psi_\eta \end{pmatrix}, \quad A = \begin{pmatrix} 0 & 1 \\ -1 & 0 \end{pmatrix}, \quad G(\xi,\eta) = \begin{pmatrix} P_f\,\eta & P_f\,\xi \\ 0 & 0 \end{pmatrix},$$

A.2. Oseen flow

and write (A.51) as a first order system:

$$\vec{v}_\xi + A\vec{v}_\eta + G(\xi,\eta)\vec{\mu} = 0. \tag{A.52}$$

Expanding $\vec{v}$ in the same manner as above,

$$\vec{v} = \beta^{(1)}\vec{r}_1 + \varepsilon\beta^{(2)}\vec{r}_2 \tag{A.53}$$

and doing the scale transformation $(\xi,\eta) \to (\varepsilon\xi, \varepsilon\eta)$ again, we have to keep in mind that $\lambda^{(2)}$ actually equals $\varepsilon\lambda^{(2)}$,

$$\left(\beta_\xi^{(1)} + i\beta_\eta^{(1)}\right)\vec{r}_1 + \varepsilon\left(\beta_\xi^{(2)} - i\beta_\eta^{(2)}\right)\vec{r}_2 + \varepsilon G(\varepsilon\xi, \varepsilon\eta)\lambda^{(1)}\vec{r}_1 = 0. \tag{A.54}$$

The next step is the projection part of the method,

$$\beta_\xi^{(1)} + i\beta_\eta^{(1)} = -\frac{i}{2}P_f(\xi - i\eta)\lambda^{(1)}, \tag{A.55a}$$

$$\beta_\xi^{(2)} - i\beta_\eta^{(2)} = \frac{i}{2}P_f(\xi - i\eta)\lambda^{(1)}. \tag{A.55b}$$

These are transformed to characteristic coordinates

$$\beta_s^{(1)} = -\frac{P_f}{2}i(2s - i(1+i\tau))c_8(\tau)e^{\tilde{a}^2[s^2-is(1+i\tau)]} \tag{A.56a}$$

$$\beta_{\bar{s}}^{(2)} = \frac{P_f}{2}(1+i\bar{\tau})c_8(\bar{\tau}+2\bar{s})e^{\tilde{a}^2[-\bar{s}^2+i\bar{s}(1+i\bar{\tau})]} \tag{A.56b}$$

and integrated,

$$\beta^{(1)} = -iPr\, c_8(\tau)e^{\tilde{a}^2[s^2-is(1+i\tau)]}, \tag{A.57a}$$

$$\beta^{(2)} = \frac{P_f}{2}(1+i\bar{\tau})\int_0^{\bar{s}} c_8(\bar{\tau}+2\omega)e^{\tilde{a}^2[-\omega^2+i\omega(1+i\bar{\tau})]}d\omega + c_{10}(\bar{\tau}). \tag{A.57b}$$

We can write the boundary conditions (3.47a) and (3.47b) in the following form:

$$(1+ih')\beta^{(1)} - (1-ih')\beta^{(2)} = -iP_f\xi\tilde{a}hh' \tag{A.58a}$$

185

Appendix A. Auxiliary calculations

$$(1+ih')\beta^{(1)} + (1-ih')\beta^{(2)} = -P_f\xi\tilde{a}h \qquad (A.58b)$$

$$(A.58a) + (A.58b) \Rightarrow \quad \beta^{(1)}\Big|_{s=0} = -\frac{P_f}{2}\tau\tilde{a}h(\tau) \qquad (A.59a)$$

$$(A.58b) - (A.58a) \Rightarrow \quad \beta^{(2)}\Big|_{\bar{s}=0} = -\frac{P_f}{2}\bar{\tau}\tilde{a}h(\bar{\tau}) \,. \qquad (A.59b)$$

We apply these to $\beta^{(1),(2)}$:

$$c_8(\tau) = -\frac{iP_f}{2Pr}\tau\tilde{a}h(\tau), \quad c_{10}(\bar{\tau}) = -\frac{1}{2}P_f\bar{\tau}\tilde{a}h(\bar{\tau}) \,.$$

This gives (3.50a) and (3.50b).

One has to decompose and solve another Laplace equation for the first order correction to the potential function φ:

$$\varphi_\xi + i\varphi_\eta = 0 \,,$$
$$\varphi_\xi - i\varphi_\eta = 0 \,.$$

The resulting functions depend only on τ and $\bar{\tau}$ respectively. The results would be added to $\psi_{\xi,\eta}$ according to (3.45a) and (3.45b):

$$\psi_\xi = -i\left(\beta^{(1)}(-\bar{s}, \bar{\tau}+2\bar{s}) - \beta^{(2)}(\bar{s},\bar{\tau})\right) + c_{11}(\bar{\tau}+2\bar{s}) + c_{12}(\bar{\tau}) \,,$$
$$\psi_\eta = \beta^{(1)}(-\bar{s}, \bar{\tau}+2\bar{s}) + \beta^{(2)}(\bar{s},\bar{\tau}) - i\left(c_{11}(\bar{\tau}+2\bar{s}) - c_{12}(\bar{\tau})\right) \,.$$

But this is already included in the integration constants c_8 and c_{10} of $\beta^{(1),(2)}$ determined from the corresponding boundary conditions (A.59a) depending on $\tau = \bar{\tau}+2\bar{s}$ and (A.59b) depending on $\bar{\tau}$. Hence, the contribution from the φ-solution is automatically included in $\beta^{(1),(2)}$. It happens because (3.45a) and (3.45b) had been differentiated once to get (3.46a).

At $\eta \to \infty \Leftrightarrow s \to -i\infty, \bar{s} \to i\infty$, we demand $\beta^{(2)}$ to be analytic and it is easy to see that $\beta^{(1)}$ vanishes in this limit. Thus, $\psi_{\xi,\eta}$ also vanish in this limit. The far field behaviour of the corrected stream function $\psi_{\xi,\eta} + \psi_{\xi,\eta}^{\text{Iv}}$ remains correct. This would not happen if c_{11}, c_{12} were different from zero. It shows again that these contributions should be dropped. We found a fully consistent first order correction to the flow velocity field fulfilling the interface conditions and leaving the far field conditions untouched:

$$\psi_\xi = -\mathrm{i}\left(\beta^{(1)} - \beta^{(2)}\right), \qquad \text{(A.60a)}$$

$$\psi_\eta = \beta^{(1)} + \beta^{(2)}. \qquad \text{(A.60b)}$$

It will turn out that $\beta^{(2)}$ does not have any effect on our theory because it is actually $\mathcal{O}(\varepsilon)$ just as $\lambda^{(2)}$.

Calculation of an asymptotically consistent correction to the temperature field

We want to derive decomposed equations, which can be solved analytically. But since (3.44a) is nonlinear, there is no guarantee for success. At first, we write the equation as a first order system:

$$\vec{\vartheta}_\xi + A\vec{\vartheta}_\eta + B\vec{\vartheta} = -\mathrm{i}\left(\beta^{(1)} - \beta^{(2)}\right) \mathrm{e}^{-I_1(\eta)} \vec{j}. \qquad \text{(A.61)}$$

Here, $\vec{j}$ is a basis unit vector in the abstract Zauderer decomposition space referring to the upper component, i.e. $\vec{r}_{1,2} \cdot \vec{j} = \mp \mathrm{i}$ and $\vec{\vartheta} = T_\xi \vec{j} + T_\eta \vec{k}$. We have used the following quantities:

$$\vec{\vartheta} = \begin{pmatrix} T_\xi \\ T_\eta \end{pmatrix}, \quad A = \begin{pmatrix} 0 & 1 \\ -1 & 0 \end{pmatrix}, \quad B = \begin{pmatrix} -\psi_\eta - \psi_\eta^{\text{Iv}} & \psi_\xi + \psi_\xi^{\text{Iv}} \\ 0 & 0 \end{pmatrix}.$$

Similar to the procedure when solving systems of ordinary differential equations, we expand the solution in the liquid in eigenvectors of A:

$$\vec{\vartheta} = M\vec{r}_1 + \varepsilon N\vec{r}_2.$$

In the solid, we can write

$$\vec{\vartheta}^s = N^s \vec{r}_2$$

Appendix A. Auxiliary calculations

just as in (2.33a), because in the limit $P_c \to 0$, the right hand side of (3.44b) is zero. For arbitrary P_c, the $\vec{r}_1$-term in $\vec{\vartheta}^s$ would have to be taken into account in order to obtain a consistent solution, and things would be much more complicated. We use the stretched scale again. In this respect, we remember $\beta^{(2)} \to \varepsilon \beta^{(2)}$,

$$(M_\xi + \mathrm{i}M_\eta)\,\vec{r}_1 + \varepsilon \left(N_\xi - \mathrm{i}N_\eta\right) \vec{r}_2 + \varepsilon B M \vec{r}_1 = -\mathrm{i}\varepsilon \beta^{(1)} \mathrm{e}^{-I_1(\eta)} \vec{j}. \quad \text{(A.62)}$$

This is projected onto the invariant subspaces of A. The characteristic coordinates are given in (3.15a) and (3.15b). The decomposed equations are (3.52a) and (3.52b). a_1 is a function determined by calculating the projection of $B\vec{r}_1$ on $\vec{r}_1$ explicitly, i.e. $P_1 B \vec{r}_1 = -a_1 \vec{r}_1$. The boundary conditions are (3.16a) and (3.16b). Let us explain the formation of (3.53a) a bit. The homogenous solution is just

$$M^h(s,\tau) = c_{13} \exp\left[\int_0^s a_1(\omega,\tau)\,\mathrm{d}\omega\right]$$

with

$$c_{13} = \frac{\mathrm{i}\left[(1+\mathrm{i}\tau)\,h(\tau)\right]'}{2\,(1+\mathrm{i}h'(\tau))}$$

gained from the boundary condition (3.16a). $\beta^{(1)}$ had already been calculated in the right coordinates (s,τ) above, see equation (3.50a). For the particular part M^p of the solution M, we start to vary the constant: $M^p = \frac{c_{13}(s)}{c_{13}} M^h$,

$$c'_{13}(s) = \beta^{(1)}(s,\tau) \exp\left(-I_1(\eta) - \int_0^s a_1(\omega,\tau)\,\mathrm{d}\omega\right)$$

$$= \beta^{(1)}(s,\tau) \exp\left(-\int_0^s \underbrace{\left(\mathrm{i}\psi_\xi^{\mathrm{Iv}}(1+\mathrm{i}\omega) - \frac{1}{2}\psi_\eta^{\mathrm{Iv}}(s+\tau,1+\mathrm{i}s) + \frac{\mathrm{i}}{2}\psi_\xi^{\mathrm{Iv}}(1+\mathrm{i}s)\right)}_{=a_2(\omega,\tau)}\,\mathrm{d}\omega\right)$$

$$= -\frac{P_f}{4}\tilde{a}\tau h(\tau) \exp\left(\tilde{\alpha}^2\left(s^2 - \mathrm{i}s\,(1+\mathrm{i}\tau)\right) - \int_0^s a_2(\omega,\tau)\,\mathrm{d}\omega\right).$$

A.2.3. WKB solution

Proof that (3.21) is valid for arbitrary flows

We start from equations (3.52a)-(3.52b) determining the Zauderer coefficients M and N respectively. These equations result from the application of the Zauderer decomposition scheme in subsection A.2.2 to the expanded diffusion-advection equation (3.44a) for the case of an Oseen flow approximation. However, when taking a closer look at the equations together with the definition (3.51) of the function $a_1(s,\tau)$, it can be seen clearly that no particular form of $\psi_{\xi,\eta}$ or $\psi_{\xi,\eta}^{\text{Iv}}$ has been inserted up to this point. Regarding equations (A.60a) and (3.41), one may take one step back and replace the inhomogeneities $\pm\frac{1}{2}\beta^{(1)}(s,\tau)e^{-I_1(\eta)}$ by $\mp\frac{i}{2}\psi_\xi T_\eta^{\text{Iv}}$ in order to make the expressions even more general. These functions are abbreviated as

$$j_o(s,\tau) = -\frac{i}{2}\psi_\xi(\tau+s, 1+is)\, T_\eta^{\text{Iv}}(1+is)\,. \tag{A.63}$$

So within the convective symmetric model used here, the Zauderer decomposition scheme will generally lead to

$$M_s = a_1(s,\tau)M + j_o(s,\tau) \tag{A.64a}$$
$$N_{\bar{s}} = -a_1(-\bar{s}, \bar{\tau}+2\bar{s})M(-\bar{s},\bar{\tau}+2\bar{s}) - j_o(-\bar{s},\bar{\tau}+2\bar{s}) \tag{A.64b}$$

with the precondition that the similarity solutions T^{Iv} and ψ^{Iv} for the case without surface tension exist. The solutions to these equations are

$$M = g_o(s,\tau)\left[\frac{i\left[(1+i\tau)h(\tau)\right]'}{2(1+ih'(\tau))} + \int_0^s \frac{j_o(\omega,\tau)}{g_o(\omega,\tau)}d\omega\right] \tag{A.65a}$$

Appendix A. Auxiliary calculations

$$N_{\bar{s}} = -\int_0^{\bar{s}} [a_1(-\omega, \bar{\tau} + 2\omega) M(-\omega, \bar{\tau} + 2\omega) \qquad \text{(A.65b)}$$
$$-j_o(-\omega, \bar{\tau} + 2\omega)] d\omega + \frac{\tilde{F}'(\bar{\tau})}{1 - ih'(\bar{\tau})}$$

with

$$g_o(s, \tau) = \exp\left(\int_0^s a_1(\omega, \tau) d\omega\right) \qquad \text{(A.66)}$$

and $\tilde{F}$ given in (3.19). To fulfill the far field boundary condition (3.10d), we need $N(\bar{s} \to i\infty, \bar{\tau}) \to 0$. In addition to that, rename $\bar{\tau} \to \xi$, which is now the continuation of the original parabolic coordinate ξ to the complex plane, and we substitute $\omega = \frac{1}{2}(\xi' - \xi)$ in the integrals:

$$\frac{\tilde{F}'(\xi)}{1 - ih'(\xi)} = i \int_{i\infty}^{\xi} \left[a_1\left(\frac{1}{2}(\xi - \xi'), \xi'\right) M\left(\frac{1}{2}(\xi - \xi'), \xi'\right) \right.$$
$$\left. - j_o\left(\frac{1}{2}(\xi - \xi'), \xi'\right) \right] d\omega . \qquad \text{(A.67)}$$

Now we calculate

$$\frac{\partial}{\partial \xi} M\left(\frac{1}{2}(\xi - \xi'), \xi'\right) = \frac{1}{2} a_1\left(\frac{1}{2}(\xi - \xi'), \xi'\right) M\left(\frac{1}{2}(\xi - \xi'), \xi'\right) ,$$
$$+ \frac{1}{2} j_o\left(\frac{1}{2}(\xi - \xi'), \xi'\right) .$$
$$\text{(A.68)}$$

This can be seen easily from the form (A.65a) of M and we used

$$\frac{\partial}{\partial \xi} g_o\left(\frac{1}{2}(\xi - \xi'), \xi'\right) = a_1\left(\frac{1}{2}(\xi - \xi'), \xi'\right) g_o\left(\frac{1}{2}(\xi - \xi'), \xi'\right) .$$

Thus, $a_1 M$ in (A.67) can be expressed by its own derivative with respect to ξ using (A.68):

$$\frac{\tilde{F}'(\xi)}{1-\mathrm{i}h'(\xi)} = 2\mathrm{i} \int_{\mathrm{i}\infty}^{\xi} \left[\frac{\partial}{\partial \xi} M\left(\frac{1}{2}(\xi-\xi'),\xi'\right) + j_o\left(\frac{1}{2}(\xi-\xi'),\xi'\right) \right.$$
$$\left. - j_o\left(\frac{1}{2}(\xi-\xi'),\xi'\right) \right] \mathrm{d}\omega$$
$$= 2\mathrm{i} \int_{\mathrm{i}\infty}^{\xi} \left[\frac{\partial}{\partial \xi} M\left(\frac{1}{2}(\xi-\xi'),\xi'\right) \right] \mathrm{d}\omega$$
$$= 2\mathrm{i} \frac{\partial}{\partial \xi} \int_{\mathrm{i}\infty}^{\xi} \left[M\left(\frac{1}{2}(\xi-\xi'),\xi'\right) \right] \mathrm{d}\omega + \frac{[(1+\mathrm{i}\xi)\,h(\xi)]'}{1+\mathrm{i}h'(\xi)}.$$
(A.69)

Multiplying both sides of (A.69) with $1 - \mathrm{i}h'(\xi)$ and integrating by parts yields (3.21) **q. e. d.**

Now that the proof was carried out, we may continue with the WKB solution. The first integral on the right hand side of (3.56) is integrated by parts,

$$\sigma\kappa(\xi) = (1-\mathrm{i}\xi)\,h(\xi) + \int_{\mathrm{i}\infty}^{\xi} (1+\mathrm{i}\xi')h(\xi')\,\mathrm{e}^{I_3(-\bar{s}',\xi')}\frac{\mathrm{d}I_3(-\bar{s}',\xi')}{\mathrm{d}\xi'}\,\mathrm{d}\xi'$$
$$-\frac{\mathrm{i}}{2}P_f\tilde{a}\int_{\mathrm{i}\infty}^{\xi} \xi'h(\xi')\,\mathrm{e}^{I_3(-\bar{s}',\xi')} \int_{0}^{\frac{1}{2}(\xi-\xi')} \mathrm{e}^{\tilde{a}^2\left(\omega^2-\mathrm{i}\omega(1+\mathrm{i}\xi')\right)}$$
$$\times \exp\left(-\int_{0}^{\omega} a_2(\omega',\xi')\,\mathrm{d}\omega'\right) \mathrm{d}\omega\,\mathrm{d}\xi'.$$
(A.70)

Appendix A. Auxiliary calculations

Explicit expressions for $I_3(-\bar{s}', \xi')$ and its total derivative with respect to ξ' are needed.

Calculation of I_3

We start from (3.56). The first integral on the right hand side is tackled with integration by parts. This seems reasonable because one factor appears as a derivative. But we need to calculate $I_3\left(\frac{1}{2}(\xi - \xi'), \xi'\right)$. These calculations are quite extensive. It is similar to the evaltuation of $I_1(\eta)$. We must pay attention and keep in mind that the expressions should not get out of hand. It is possible but we always have to choose the same convenient factorization, i.e. we collect terms of $P_f \tilde{a}/Re$ and terms containing erfc. The integral is

$$I_3(-\bar{s}', \xi') = \frac{1}{2} \int_0^{\frac{1}{2}(\xi - \xi')} \left[\psi_\eta^{\text{Iv}}(\omega + \xi', 1 + i\omega) - i\psi_\xi^{\text{Iv}}(1 + i\omega) \right] d\omega$$

and it is transformed back: $\eta' = 1 + i\omega$. Integration is less extensive with this integration variable:

$$I_3(\eta, \xi') = -\frac{i}{2} \int_1^\eta \left[\psi_\eta^{\text{Iv}}\left(\xi' - i(\eta' - 1), \eta'\right) - i\psi_\xi^{\text{Iv}}\left(\eta'\right) \right] d\eta'$$
$$= -iI_3(-\bar{s}', \xi').$$

Note, that $I_3(-\bar{s}', \xi')$ and $I_3(\eta, \xi')$ are actually different functions. The one is not gained from the other by just replacing arguments but by substitution under the integral. The bracket expressions after the identifier "I_3" are more like a part of the name than arguments of the function. If one considers them as arguments, one will always have to know, which version of the integral is viewed. $I_3(-\bar{s}', \xi')$ can be calculated by replacing $\eta = 1 + i\omega = 1 + \frac{i}{2}(\xi - \xi')$ in $I_3(\eta, \xi')$. The same applies for the other I_k presented in this work. The notation is a bit tricky but reasonable. The (η, ξ')-notation is used for evaluation

A.2. Oseen flow

of the integrals. It makes the occuring expressions more convenient. The $(-\bar{s}', \xi')$-notation is the one, which has to be used in the upcoming sections and chapters. Using

$$I_4(\eta, \xi') = \int_1^\eta \psi_\eta^{\text{Iv}}\left(\xi' - \mathrm{i}(\eta' - 1), \eta'\right) \mathrm{d}\eta',$$

we can write

$$I_3(\eta, \xi') = -\frac{\mathrm{i}}{2} I_4(\eta, \xi') - \frac{1}{2} I_1(\eta). \qquad (\text{A.71})$$

We could also set

$$I_2(\eta, \xi') = \mathrm{i} \int_1^\eta a_2\left(-\mathrm{i}(\eta - 1), \xi'\right) \mathrm{d}\eta'$$

and write

$$I_3(\eta, \xi') = -I_2(\eta, \xi') - I_1(\eta, \xi'). \qquad (\text{A.72})$$

Both ways (A.71) and (A.72) have been taken to compare and verify results. This is advisable, because $I_3(\eta, \xi')$ is an elementary part of the results in subsection 3.2.3 and the theory behind it, but its calculation is prone to mistakes. We will only present the path of (A.71), because it is shorter. Later on, we will just give the result for $I_2(\eta, \xi')$, which the reader may verify herself or himself and use it to evaluate (A.72).

For the first path, we need to evaluate only $I_4(\eta, \xi')$ because $I_1(\eta)$ is already known. The Terms are integrated in the same manner as during the calculation of $I_1(\eta)$:

$$I_4(\eta, \xi') = \int_1^\eta \psi_\eta^{\text{Iv}}\left(\xi' - \mathrm{i}(\eta' - 1), \eta'\right) \mathrm{d}\eta'$$

$$= \int_1^\eta \left(\xi' - \mathrm{i}(\eta' - 1)\right) P_f \left(1 - \frac{\mathrm{erfc}\,(\tilde{\alpha}\eta')}{\mathrm{erfc}\,(\tilde{\alpha})}\right) \mathrm{d}\eta'$$

Appendix A. Auxiliary calculations

$$
\begin{aligned}
&= \mathrm{i}\left(1 - \mathrm{i}\xi'\right) \eta P_f \left(1 - \frac{\operatorname{erfc}\left(\tilde{\alpha}\eta\right)}{\operatorname{erfc}\left(\tilde{\alpha}\right)}\right) \\
&\quad - \frac{\mathrm{i}}{2}\eta^2 P_f \left(1 - \frac{\operatorname{erfc}\left(\tilde{\alpha}\eta\right)}{\operatorname{erfc}\left(\tilde{\alpha}\right)}\right) \\
&\quad + \underbrace{\int_1^\eta \left[\frac{\mathrm{i}}{2}\eta'^2 - \mathrm{i}\left(1 - \mathrm{i}\xi'\right)\eta'\right] P_f \tilde{\alpha} e^{\tilde{\alpha}^2\left(1 - \eta'^2\right)} \mathrm{d}\eta'}_{=t_4}.
\end{aligned}
$$

We continue with auxiliary term t_4:

$$
\begin{aligned}
t_4 &= -\left.\frac{\mathrm{i}P_f \tilde{\alpha}}{4\tilde{\alpha}^2}\eta' e^{\tilde{\alpha}^2\left(1 - \eta'^2\right)}\right|_1^\eta \\
&\quad + \frac{\mathrm{i}P_f \tilde{\alpha}}{4\tilde{\alpha}^2} \int_1^\eta e^{\tilde{\alpha}^2\left(1 - \eta'^2\right)} \mathrm{d}\eta' + \mathrm{i}\left(1 - \mathrm{i}\xi'\right) \left.\frac{P_f \tilde{\alpha}}{2\tilde{\alpha}^2} e^{\tilde{\alpha}^2\left(1 - \eta'^2\right)}\right|_1^\eta \\
&= \frac{\mathrm{i}P_f}{2Re}\left[1 - \frac{\operatorname{erfc}\left(\tilde{\alpha}\eta\right)}{\operatorname{erfc}\left(\tilde{\alpha}\right)}\right] \\
&\quad + \frac{\mathrm{i}P_f \tilde{\alpha}}{Re}\left[\frac{1}{2}\left(1 - \eta e^{\tilde{\alpha}^2\left(1 - \eta^2\right)}\right) + \left(1 - \mathrm{i}\xi'\right)\left(e^{\tilde{\alpha}^2\left(1 - \eta^2\right)} - 1\right)\right].
\end{aligned}
$$

Now we can write $I_4(\eta, \xi')$:

$$
\begin{aligned}
I_4(\eta, \xi') &= \mathrm{i}P_f \left[1 - \frac{\operatorname{erfc}\left(\tilde{\alpha}\eta\right)}{\operatorname{erfc}\left(\tilde{\alpha}\right)}\right]\left[\eta\left(1 - \mathrm{i}\xi'\right) - \frac{\eta^2}{2} + \frac{1}{2Re}\right] \\
&\quad + \frac{\mathrm{i}P_f \tilde{\alpha}}{Re}\left[\frac{1}{2}\left(1 - \eta e^{\tilde{\alpha}^2\left(1 - \eta^2\right)}\right) + \left(1 - \mathrm{i}\xi'\right)\left(e^{\tilde{\alpha}^2\left(1 - \eta^2\right)} - 1\right)\right].
\end{aligned}
$$

Now we can write $I_3(\eta, \xi')$:

$$
\begin{aligned}
I_3(\eta, \xi') &= -\frac{\mathrm{i}}{2}I_4(\eta, \xi') - \frac{1}{2}I_1(\eta) \\
&= \frac{P_f}{2}\left[1 - \frac{\operatorname{erfc}\left(\tilde{\alpha}\eta\right)}{\operatorname{erfc}\left(\tilde{\alpha}\right)}\right]\left[\eta\left(1 - \mathrm{i}\xi'\right) - \eta^2\right] \\
&\quad + \frac{P_f \tilde{\alpha}}{2Re}\left[1 - e^{\tilde{\alpha}^2\left(1 - \eta^2\right)}\right]\left[\eta - \left(1 - \mathrm{i}\xi'\right)\right].
\end{aligned}
$$

A.2. Oseen flow

Now we can write $I_3\left(\frac{1}{2}(\xi - \xi'), \xi'\right)$. We have to replace $\eta = 1 + i\omega = 1 + \frac{i}{2}(\xi - \xi')$ in $I_3(\eta, \xi')$,

$$I_3(-\bar{s}', \xi') = \frac{iP_f\tilde{a}}{4Re}\left[1 - e^{\tilde{\alpha}^2\left(\frac{1}{4}(\xi-\xi')^2 - i(\xi-\xi')\right)}\right](\xi + \xi')$$
$$+ \frac{P_f}{4}\left[1 - \frac{\mathrm{erfc}\left(\tilde{\alpha}\left(1 + \frac{i}{2}(\xi - \xi')\right)\right)}{\mathrm{erfc}(\tilde{\alpha})}\right]\left[\frac{1}{2}\left(\xi^2 - \xi'^2\right) - i(\xi + \xi')\right]. \tag{A.73}$$

As mentioned above, we will give I_2 for completeness. We write it in the form $I_2(\omega, \xi')$ where $\omega = \frac{1}{2}(\xi - \xi') = -\bar{s}'$ can be identified:

$$I_2(\omega, \xi') = -\frac{P_f}{2}\left[1 - \frac{\mathrm{erfc}(\tilde{\alpha}(1 + i\omega))}{\mathrm{erfc}(\tilde{\alpha})}\right]\left[(1 + i\omega)(1 - i\xi') + \frac{1}{Re}\right]$$
$$+ \frac{P_f\tilde{a}}{2Re}\left[(1 - i\xi')\left(1 - e^{\tilde{\alpha}^2(\omega^2 - 2i\omega)}\right) + i\omega\right]. \tag{A.74}$$

We are now ready to evaluate the integral in (3.56) using integration by parts:

$$\sigma\kappa(\xi)a(\theta) = 2h(\xi) + (1 + i\xi')h(\xi')e^{I_3(-\bar{s}', \xi')}\Big|_\xi^{i\infty}$$
$$- \int_\xi^{i\infty} (1 + i\xi')h(\xi')e^{I_3(-\bar{s}', \xi')}\frac{dI_3(-\bar{s}', \xi')}{d\xi'}d\xi'$$
$$= (1 - i\xi')h(\xi') + \int_{i\infty}^\xi (1 + i\xi')h(\xi')e^{I_3(-\bar{s}', \xi')}\frac{dI_3(-\bar{s}', \xi')}{d\xi'}d\xi'.$$

We have used $I_3(0, \xi) = 0$ (in the $(-\bar{s}, \xi')$-notation) and the requirement for the function to vanish at $\xi' \to i\infty$.

Calculation of the derivative of I_3

$$\frac{dI_3(-\bar{s}', \xi')}{d\xi'} = -\frac{1}{2}a_1\left(\frac{1}{2}(\xi - \xi'), \xi'\right) + \frac{1}{2}\int_0^{\frac{1}{2}(\xi-\xi')}\frac{d\psi_\eta^{IV}(\omega + \xi', 1 + i\omega)}{d\xi'}d\omega$$

Appendix A. Auxiliary calculations

$$= -\frac{1}{4}\left[\psi_\eta^{\mathrm{Iv}}\left(\frac{1}{2}(\xi+\xi'), 1+\frac{\mathrm{i}}{2}(\xi-\xi')\right) - \mathrm{i}\psi_\xi^{\mathrm{Iv}}\left(1+\frac{\mathrm{i}}{2}(\xi-\xi')\right)\right]$$

$$+ \frac{1}{2}\int_0^{\frac{1}{2}(\xi-\xi')} \frac{\mathrm{d}\psi_\eta^{\mathrm{Iv}}(\omega+\xi', 1+\mathrm{i}\omega)}{\mathrm{d}\xi'}\mathrm{d}\omega,$$

since in the limit $P_c \to 0$

$$\frac{\mathrm{d}\psi_\eta^{\mathrm{Iv}}(\omega+\xi', 1+\mathrm{i}\omega)}{\mathrm{d}\xi'} = P_f\left(1 - \frac{\mathrm{erfc}\,(\tilde{\alpha}\,(1+\mathrm{i}\omega))}{\mathrm{erfc}\,(\tilde{\alpha})}\right),$$

we calculate the following integral:

$$\frac{P_f}{2}\int_0^{\frac{1}{2}(\xi-\xi')}\left(1 - \frac{\mathrm{erfc}\,(\tilde{\alpha}\,(1+\mathrm{i}\omega))}{\mathrm{erfc}\,(\tilde{\alpha})}\right)\mathrm{d}\omega$$

$$= \frac{P_f}{2}\omega\Big|_0^{\frac{1}{2}(\xi-\xi')} - \frac{P_f}{2}\omega\frac{\mathrm{erfc}\,(\tilde{\alpha}\,(1+\mathrm{i}\omega))}{\mathrm{erfc}\,(\tilde{\alpha})}\Big|_0^{\frac{1}{2}(\xi-\xi')} \underbrace{- \frac{\mathrm{i}P_f}{2}\tilde{a}e^{\tilde{\alpha}^2}\int_0^{\frac{1}{2}(\xi-\xi')}e^{-\tilde{\alpha}^2(1+\mathrm{i}\omega)^2}\mathrm{d}\omega}_{=t_5}$$

$$= \frac{P_f}{4}\left(1 - \frac{\mathrm{erfc}\,(\tilde{\alpha}\,(1+\frac{\mathrm{i}}{2}(\xi-\xi')))}{\mathrm{erfc}\,(\tilde{\alpha})}\right) + t_5.$$

For calculation of the auxiliary term t_5, we resubstitute $\omega = -\mathrm{i}(\eta'-1)$,

$$t_5 = \frac{\mathrm{i}P_f}{2}\tilde{a}e^{\tilde{\alpha}^2}\int_1^\eta (\eta'-1)\,e^{-\tilde{\alpha}^2\eta'^2}\mathrm{d}\eta'$$

$$= \frac{\mathrm{i}P_f\tilde{a}}{4\tilde{\alpha}^2}\,e^{\tilde{\alpha}^2(1-\eta'^2)}\Big|_\eta^1 - \frac{\mathrm{i}P_f}{2}\tilde{a}e^{\tilde{\alpha}^2}\int_1^\eta e^{-\tilde{\alpha}^2\eta'^2}\mathrm{d}\eta'$$

$$= \frac{\mathrm{i}P_f\tilde{a}}{2\mathrm{Re}}\left[1 - e^{\tilde{\alpha}^2(1-\eta^2)}\right] - \frac{\mathrm{i}P_f}{2}\left[1 - \frac{\mathrm{erfc}\,(\tilde{\alpha}\eta)}{\mathrm{erfc}\,(\tilde{\alpha})}\right].$$

Reinserting $\eta = 1 + \frac{1}{2}(\xi-\xi')$ into t_5, we are ready to collect the

A.2. Oseen flow

corresponding terms in $\frac{dI_3(-\bar{s}',\xi')}{d\xi'}$:

$$\frac{dI_3(-\bar{s}',\xi')}{d\xi'} = -\frac{iP_f}{4}(1 - i\xi')\left[1 - \frac{\operatorname{erfc}\left(\tilde{\alpha}\left(1 + \frac{i}{2}(\xi - \xi')\right)\right)}{\operatorname{erfc}(\tilde{\alpha})}\right]$$
$$+ \frac{iP_f\tilde{\alpha}}{4Re}\left[1 - e^{\tilde{\alpha}^2\left(\frac{1}{4}(\xi-\xi')^2 - \frac{1}{2}(\xi-\xi')\right)}\right]. \quad (A.75)$$

Of course, this result can be approved by direct derivation of (A.73). Just the sequence of differentiation and integration must be swapped. One should get the same, and this was checked indeed. Inserting (A.75) into (A.70), one ends up with (3.57).

Finding the WKB solution (3.24)

We use the ansatz (A.19) in (3.58),

$$-\sigma\left(\frac{h''(\xi)}{\sqrt{1+\xi^2}} + \frac{\xi h'(\xi)}{(1+\xi^2)^{\frac{3}{2}}}\right) = (1 - i\xi)\,h(\xi). \quad (A.76)$$

We have used $I_3(0, \xi) = 0$ again. The derivatives of $h(\xi)$ can be expressed in terms of $h(\xi)$ itself. We pay attention only to the two lowest orders:

$$h'(\xi) = \left(\frac{S_0'(\xi)}{\varepsilon} + S_1'(\xi)\right) h(\xi),$$
$$h''(\xi) = \left(\frac{S_0''(\xi)}{\varepsilon} + S_1''(\xi) + \frac{S_0'^2(\xi)}{\varepsilon^2} + S_1'^2(\xi) + \frac{2}{\varepsilon}S_0'(\xi)S_1'(\xi)\right) h(\xi).$$

Now we can set up (A.76) up to the order ε^1:

$$0 = S_0'(\xi)^2 + \sqrt{1+\xi^2}\,(1 - i\xi)$$
$$+ \varepsilon\left(\frac{\xi}{1+\xi^2}S_0'(\xi) + 2S_0'(\xi)S_1'(\xi) + S_0''(\xi)\right). \quad (A.77)$$

The zero order is given by (3.25), calculated in appendix A.1.3. The first order can now be calculated:

$$\varepsilon^1: \quad S_1'(\xi) = -\frac{1}{2}\frac{\xi}{1+\xi^2} - \frac{1}{2}\frac{d}{d\xi}\left[\ln\left(S_0'(\xi)\right)\right]$$

Appendix A. Auxiliary calculations

$$S_1(\xi) = \ln\left((1+i\xi)^{-\frac{3}{8}}(1-i\xi)^{-\frac{5}{8}}\right).$$

In the calculation of the first order, the integration constants are chosen properly. The important issue is that the asymptotic matching works later on.

A.2.4. Local equation

We use the simple formulas (A.22a)-(A.22c) and $d\xi = i\sigma^\alpha dt$, which can be complemented by

$$(1+i\xi)\,h(\xi) = (2-\sigma^\alpha t)\,\sigma^\alpha \phi(t)\,. \qquad (A.78)$$

We have to transform

1. $a_1\left(\frac{1}{2}(\xi'-\xi''),\xi''\right)$
2. $I_3\left(\frac{1}{2}(\xi'-\xi''),\xi''\right)$
3. $M\left(\frac{1}{2}(\xi'-\xi''),\xi''\right)$
4. $I_1\left(\frac{1}{2}(\xi'-\xi''),\xi''\right)$
5. auxiliary term t_6 from equation (3.60)

and we are going to proceed in that very order. Because of $d\xi = i\sigma^\alpha dt$ and accordingly $d\xi''d\xi' = -\sigma^{2\alpha}dt''dt'$, even a zero order contribution from the above listed terms will end up as $\mathcal{O}\left(\sigma^{2\alpha}\right)$. Thus, we calculate almost everything only up to the order of $\sigma^{2\alpha}$. This should be by far enough to perform the asymptotic matching to the WKB solution (3.59) afterwards. When products are formed, for example $a_1 M$ occurs, more terms will drop out.

Transformation of a_1
We write $a_1\left(\frac{1}{2}(\xi'-\xi''),\xi''\right)$ explicitly using its definition (3.51). This has not been done so far. The derivatives $\psi_{\xi\eta}$ of the stream function

are known,

$$\begin{aligned}a_1\left(\frac{1}{2}(\xi'-\xi''),\xi''\right) &= -\frac{\mathrm{i}P_f}{2Re}\tilde{a}\left(\mathrm{e}^{\tilde{\alpha}^2\left[\frac{1}{4}(\xi'-\xi'')^2-\mathrm{i}(\xi'-\xi'')\right]}-1\right)\\&-\frac{\mathrm{i}}{2}P_f\left(1+\mathrm{i}\xi'\right)\left(1-\frac{\operatorname{erfc}\left(\tilde{\alpha}\left(1+\frac{\mathrm{i}}{2}(\xi'-\xi'')\right)\right)}{\operatorname{erfc}\left(\tilde{\alpha}\right)}\right).\end{aligned} \qquad (A.79)$$

Since

$$\xi'-\xi'' = \mathrm{i}\sigma^\alpha\left(t'-t''\right) = 2\mathrm{i}d_1,$$

we have to expand the following complementary error function in a *Taylor series*, because it occurs in $a_1\left(\frac{1}{2}(\xi'-\xi''),\xi''\right)$:

$$\begin{aligned}\operatorname{erfc}\left(\tilde{\alpha}\left(1+\frac{\mathrm{i}}{2}(\xi'-\xi'')\right)\right) &= \operatorname{erfc}\left(\tilde{\alpha}\left(1-d_1\right)\right)|_{d_1=0}\\&\approx \operatorname{erfc}\left(\tilde{\alpha}\right)\left[1+\tilde{a}d_1+\tilde{a}\tilde{\alpha}^2 d_1^2\right]\\&=\operatorname{erfc}\left(\tilde{\alpha}\right)\left[1+\frac{1}{2}\tilde{a}\sigma^\alpha\left(t'-t''\right)+\frac{1}{8}\tilde{a}Re\,\sigma^{2\alpha}\left(t'-t''\right)^2\right].\end{aligned} \qquad (A.80)$$

The expansion is reasonable. For example the estimation $\sqrt{\frac{Re}{2}}\sigma^\alpha \approx 2.36\times 10^{-4} \ll 1$ can be made for pivalic acid. Since

$$\frac{1}{4}(\xi'-\xi'')^2 - \mathrm{i}(\xi'-\xi'') = \sigma^\alpha(t'-t'') - \frac{\sigma^{2\alpha}}{4}(t'-t'')^2 = 2d_1 - d_1^2,$$

we have to expand the following exponential function in a Taylor series, because it occurs in $a_1\left(\frac{1}{2}(\xi'-\xi''),\xi''\right)$:

$$\begin{aligned}\mathrm{e}^{\tilde{\alpha}^2\left[\frac{1}{4}(\xi'-\xi'')^2-\mathrm{i}(\xi'-\xi'')\right]} &= \mathrm{e}^{\tilde{\alpha}^2\left(2d_1-d_1^2\right)}\\&\approx 1 + Re\,d_1 + \frac{1}{2}Re\left(Re-1\right)d_1^2\\&= 1 + \frac{1}{2}Re\,\sigma^\alpha\left(t'-t''\right) + \frac{1}{8}Re\left(Re-1\right)\sigma^{2\alpha}\left(t'-t''\right)^2.\end{aligned} \qquad (A.81)$$

Appendix A. Auxiliary calculations

We are now ready to finish the transformation of a_1:

$$a_1 = -\frac{i}{2}\left(2 - \sigma^\alpha t'\right) P_f \tilde{a} \left(\frac{1}{2}\sigma^\alpha (t' - t'') + \frac{Re}{8}\sigma^{2\alpha} (t' - t'')^2\right)$$

$$- \frac{iP_f}{2}\tilde{a}\left[\frac{1}{2}\sigma^\alpha (t' - t'') + \frac{1}{8}(Re - 1)\sigma^{2\alpha}(t' - t'')^2\right]$$

$$a_1 = \frac{i}{4}P_f \tilde{a}\sigma^\alpha (t' - t'')$$

$$- i\sigma^{2\alpha}\left[\frac{P_f}{4}\tilde{a}t'(t' - t'') - \frac{P_f}{16}\tilde{a}(t' - t'')^2 (Re + 1)\right]. \tag{A.82}$$

Of course, further terms of higher order arising from multiplication have been neglected. The reader may take more intermediate steps when verifying this result.

Transformation of I_3

We already expanded the corresponding exponential function and the corresponding complementary error function in the above paragraph. We have

$$\xi' + \xi'' = -i\left(2 - \sigma^\alpha (t' + t'')\right)$$

$$\xi'^2 - \xi''^2 = 2\sigma^\alpha (t' - t'') - \frac{1}{2}\sigma^{2\alpha}(t'^2 - t''^2)$$

$$-i(\xi' - \xi'') = \sigma^\alpha (t' - t'')$$

and we are now ready to finish the transformation of I_3:

$$I_3 = -\frac{P_f}{4}\tilde{a}\left[-2 + 2\sigma^\alpha t' - \frac{1}{2}\sigma^{2\alpha}(t'^2 - t''^2)\right]$$

$$\times \left[\frac{1}{2}\sigma^\alpha (t' - t'') + \frac{Re}{8}\sigma^{2\alpha}(t' - t'')^2\right]$$

$$- \frac{P_f}{4}\tilde{a}\left[2 - \sigma^\alpha (t' + t'')\right]\left[\frac{1}{2}\sigma^\alpha (t' - t'') + \frac{1}{8}(Re - 1)\sigma^{2\alpha}(t' - t'')^2\right]$$

$$I_3 = -\sigma^{2\alpha}\frac{P_f}{16}\tilde{a}(t' - t'')^2. \tag{A.83}$$

I_3 is $\mathcal{O}\left(\sigma^{2\alpha}\right)$.

Transformation of M

We write $M\left(\frac{1}{2}(\xi'-\xi''),\xi''\right)$ explicitly:

$$M\left(\frac{1}{2}(\xi'-\xi''),\xi''\right) = e^{I_3(-\bar{s}'',\xi'')}\left[\frac{i}{2}\frac{[(1+i\xi'')\,h(\xi'')]'}{1+ih'(\xi'')}\right.$$
$$\left.\underbrace{-\frac{P_f}{4}\tilde{a}\xi''h(\xi'')\int_0^{\frac{1}{2}(\xi'-\xi'')}e^{\tilde{a}^2\left[\omega^2-i\omega(1+i\xi'')\right]+I_2(\omega,\xi'')}d\omega}_{=t_7}\right]. \quad \text{(A.84)}$$

The expansion of e^{I_3} is

$$e^{I_3} = 1 - \frac{P_f}{16}\tilde{a}\sigma^{2\alpha}\left(t'-t''\right)^2, \quad \text{(A.85)}$$

which will be needed. Apart from that, we need

$$[(1+i\xi'')\,h(\xi'')]' = -i\left[2\dot{\phi}(t'') - \sigma^\alpha\left(\phi(t'')+\dot{\phi}(t'')t''\right)\right]$$

and

$$1 + ih'(\xi'') = 1 + \dot{\phi}(t''). \quad \text{(A.86)}$$

In M there is another auxiliary term t_7,

$$\xi''h(\xi'') = -i\left[\sigma^\alpha\phi(t'') - \sigma^{2\alpha}t''\phi(t'')\right]. \quad \text{(A.87)}$$

Thus, we need to expand the integral only up to the order σ^α to get the full term t_7 correctly up to the order $\sigma^{2\alpha}$. The upper limit of the integral is $\frac{1}{2}d_2(t'-t'')$ with $d_2 = \sigma^\alpha$. The integral is zero for $d_2 = 0$ because the limits become equal in this case. The zero order vanishes. The first order can be written as

Appendix A. Auxiliary calculations

$$\frac{d}{dd_2}\left[\int_0^{\frac{i}{2}d_2(t'-t'')} e^{\tilde{a}^2\left[\omega^2-i\omega(2-d_2t'')\right]+I_2(\omega,-i(1-d_2t''))}d\omega\right] \quad (A.88)$$

$$= \left[\left(\int_0^{\frac{i}{2}d_2(t'-t'')} \frac{d}{dd_2}\text{"integrand"}\,d\omega\right) + \text{"integrand"}\Big|_{\omega=0}\right]_{d_2=0} \frac{i}{2}d_2(t'-t'').$$

The first term in the first order becomes zero because once again the limits of the integral become equal. In the second term, the integrand is an exponential function, the exponent of which consists of a polynom in ω and $I_2(\omega, \xi'')$. The polynom vanishes for $\omega = 0$, and it is easy to see from equation (A.74) that $I_2(0, \xi'') = 0$,

$$t_7 = -\frac{1}{8} P_f \tilde{a}\sigma^{2\alpha} (t'-t'')\,\phi(t'').$$

We prepared all the ingredients to write the transformed version of M:

$$M = \frac{1}{1+\dot\phi(t'')} \overbrace{\left[1 - \frac{P_f}{16}\tilde{a}\sigma^{2\alpha}(t'-t'')^2\right]}^{=e^{I_3}}\left[\dot\phi(t'') - \frac{1}{2}\sigma^\alpha\left(\phi(t'')+t''\dot\phi(t'')\right)\right.$$

$$\left.\underbrace{-\frac{1}{8}P_f\tilde{a}\sigma^{2\alpha}(t'-t'')\,\phi(t'')}_{=t_7} - \left(1+\dot\phi(t'')\right)\right]$$

$$M = \frac{\dot\phi(t'')}{1+\dot\phi(t'')}$$
$$-\frac{\sigma^\alpha}{1+\dot\phi(t'')}\frac{1}{2}\left(\phi(t'')+t''\dot\phi(t'')\right)$$
$$-\frac{\sigma^{2\alpha}}{1+\dot\phi(t'')}\left[\frac{P_f}{16}\tilde{a}(t'-t'')^2\,\dot\phi(t'') + \frac{P_f}{8}\tilde{a}(t'-t'')\,\phi(t'')\left(1+\dot\phi(t'')\right)\right]. \quad (A.89)$$

A.2. Oseen flow

Transformation of I_1

We use (3.42). We have

$$\eta \to 1 + \frac{i}{2}(\xi' - \xi'') = 1 - \frac{1}{2}\sigma^\alpha(t' - t'').$$

All the other functions occuring in I_1 have already been transformed,

$$\begin{aligned}I_1 = &-\frac{P_f}{2}\tilde{a}\left[\frac{1}{Re} + 1 - \sigma^\alpha(t'-t'') + \frac{1}{4}\sigma^{2\alpha}(t'-t'')^2\right]\\ &\times \left[\frac{1}{2}\sigma^\alpha(t'-t'') + \frac{Re}{8}\sigma^{2\alpha}(t'-t'')^2\right]\\ &+ \frac{P_f}{2Re}\tilde{a}\left[1 - \left(1 - \frac{1}{2}\sigma^\alpha(t'-t'')\right)\right.\\ &\left.\times \left(1 - \frac{Re}{2}\sigma^\alpha(t'-t'') - \frac{Re}{8}(Re-1)\sigma^{2\alpha}(t'-t'')^2\right)\right]\end{aligned}$$

$$I_1 \approx 0. \qquad (A.90)$$

We calculated I_1 only up to the order σ^α. We will see that this is enough because the remaining factor $i\xi''h(\xi'')$ in t_6 has no zero order (see (A.87)).

Transformation of t_6

We wish to expand the exponential function in t_6 first,

$$e^{-I_1} \approx 1 + \mathcal{O}\left(\sigma^{2\alpha}\right).$$

That's half of it,

$$\frac{1}{4}\left(\xi'^2 - \xi''^2\right) - \frac{i}{2}(\xi' - \xi'') = \sigma^\alpha(t'-t'') - \frac{1}{4}\sigma^{2\alpha}\left(t'^2 - t''^2\right).$$

Now, we expand the rest of the exponential function:

$$e^{\tilde{a}^2\left[\sigma^\alpha(t'-t'') - \frac{1}{4}\sigma^{2\alpha}\left(t'^2 - t''^2\right)\right]} \approx 1 + \frac{Re}{2}\sigma^\alpha(t'-t'').$$

Appendix A. Auxiliary calculations

We now prepared all the ingredients to write the transformed version of t_6:

$$
\begin{aligned}
t_6 &= \mathrm{i}\frac{P_f}{4}\tilde{a}\overbrace{\left[\sigma^\alpha\phi(t'') - \sigma^{2\alpha}t''\phi(t'')\right]}^{=\mathrm{i}\xi''h(\xi'')\text{ from (A.87)}}\left[1 + \frac{Re}{2}\sigma^\alpha(t'-t'')\right] \\
&\approx \mathrm{i}\frac{P_f}{4}\tilde{a}\left[\sigma^\alpha\phi(t'') - \sigma^{2\alpha}t''\phi(t'')\right]\left[1 + \frac{Re}{2}\sigma^\alpha(t'-t'')\right] \\
t_6 &= \mathrm{i}\frac{P_f}{4}\tilde{a}\left[\sigma^\alpha\phi(t'') + \sigma^{2\alpha}\phi(t'')\left(\frac{Re}{2}\sigma^\alpha(t'-t'') - t''\right)\right]. \quad (\text{A.91})
\end{aligned}
$$

Here, the second factor is (A.81) reduced to the first order. We are almost done. Regarding equation (3.60), we have

$$
\sigma^{\frac{3}{7}}\kappa - t\phi(t) = -\mathrm{i}\int_\infty^t\int^{t'}\left(1 - \dot\phi(t')\right)(a_1 M + t_6)\,\mathrm{d}t''\mathrm{d}t'.
$$

All that remains to do is to multiply a_1 and M:

$$
a_1 M = \frac{\mathrm{i}}{1+\dot\phi(t'')}\frac{P_f}{4}\tilde{a}\dot\phi(t'')(t'-t''). \quad (\text{A.92})
$$

We stopped after the first order. We can now write the "inner equation" (3.62).

Matching to the WKB solution (3.59)

For $t \to \infty$, we may reduce equation (3.62) to its dominant terms:

$$
\frac{1}{\sqrt{2t}}\frac{\mathrm{d}^3\phi}{\mathrm{d}t^3} = \dot\phi t + \phi. \quad (\text{A.93})
$$

The first derivative with respect to t of (3.62) was taken to arrive at this result. We use the standard ansatz of asymptotic analysis:

$$
\phi(t) = \mathrm{e}^{S(t)}. \quad (\text{A.94})
$$

We insert the ansatz and itterate the scheme one time. I.e. we start with the zero order and proceed until we have got the first order. The first order dominant balance reads

$$c' = -\frac{5}{8t}$$

yielding (3.63). We proceed, putting the stretching transformation into the prefactor of the exponential function in the WKB solution (3.58), keeping only leading orders:

$$(1+i\xi)^{-\frac{3}{8}}(1-i\xi)^{-\frac{5}{8}} \approx \sigma^{-\frac{5}{28}}2^{-\frac{3}{8}}t^{-\frac{5}{8}}.$$

Hence, the prefactor is reproduced. Now we calculate S_0:

$$S_0 = -\int_0^t \sigma^{\frac{3}{14}}t'^{\frac{3}{4}}2^{\frac{1}{4}}\sigma^{\frac{2}{7}}dt' = -2^{\frac{1}{4}}\frac{4}{7}t^{\frac{7}{4}}\sqrt{\sigma}.$$

Hence, the exponent matches. Considering $h(\xi) = \sigma^\alpha \phi(t)$, the constants must be related by

$$A_1 = B_1 2^{-\frac{3}{8}} \sigma^{-\frac{13}{28}}.$$

This lead directly to (3.64).

A.3. Thermal resistance in cartesian coordinates

A.3.1. Interface normal, curvature and anisotropy function in cartesian coordinates

differential line element of the 2D interface:

$$ds = \sqrt{dx^2 + dy^2} = dx\sqrt{1+y_s'^2} \tag{A.95}$$

Appendix A. Auxiliary calculations

unit tangent vector from Frenet's formula:

$$\vec{t} = \frac{d\vec{r}}{ds} = \frac{1}{\sqrt{1+y_s'^2}} \frac{d}{dx}(x\vec{e}_x + y_s\vec{e}_y) = \frac{\vec{e}_x + y_s'\vec{e}_y}{\sqrt{1+y_s'^2}} \quad (A.96)$$

unit normal vector from orthonormality relation:

$$\vec{t} \cdot \vec{n} \stackrel{!}{=} 0 \quad \Rightarrow \quad \vec{n} = \frac{\vec{e}_y - y_s'\vec{e}_x}{\sqrt{1+y_s'^2}} \quad (A.97)$$

second derivative:

$$\frac{d^2\vec{r}}{ds^2} = \frac{1}{1+y_s'^2} \frac{d}{dx}\left[\vec{e}_x + y_s'\vec{e}_y\right] = \frac{y_s''}{1+y_s'^2}\vec{e}_y \quad (A.98)$$

curvature:

$$\kappa = -\vec{n} \cdot \frac{d^2\vec{r}}{ds^2} = -\frac{y_s''}{(1+y_s'^2)^{3/2}} \quad (A.99)$$

anisotropy function:

$$a(\theta) = 1 - \beta(1 - 8\cos^2\theta \sin^2\theta) \quad (A.100)$$

$$\cos\theta = \vec{n} \cdot \vec{e}_y = \frac{1}{\sqrt{1+y_s'^2}} \quad (A.101)$$

$$\sin\theta = \vec{n} \cdot \vec{e}_x = \frac{-y_s'}{\sqrt{1+y_s'^2}} \quad (A.102)$$

$$a(\theta) = 1 - \beta\left(1 - \frac{8y_s'^2}{(1+y_s'^2)^2}\right) \quad (A.103)$$

In the case of a twofold anisotropy function of capillary effects, one finds

$$\begin{aligned}a(\theta) &= 1 - \frac{\beta_2}{2}\cos(2\theta) = 1 - \frac{\beta_2}{2}\left(\cos^2\theta - \sin^2\theta\right) \\ &= 1 + \frac{\beta_2}{2}(1 - 2\cos^2\theta) = 1 + \frac{\beta_2}{2}\left(1 - \frac{2}{1+y_s'^2}\right).\end{aligned} \quad (A.104)$$

A.3.2. Derivation of the cartesian shape equation (4.43)

We want to derive equation (4.43) from the interface conditions (4.28a)-(4.28c). The conditions need to be expanded for

$$y_s = \frac{1}{2}(1 - x^2) + \zeta(x) = y_s^{\text{Iv}} + \zeta(x), \qquad \text{(A.105a)}$$
$$T \to T + T^{\text{Iv}}. \qquad \text{(A.105b)}$$

We need approximate expressions of the Ivantsov-solution $T^{l,\text{Iv}}$ and its derivatives $T_x^{l,\text{Iv}}$, $T_y^{l,\text{Iv}}$ at y_s. For convenience, we wish to work with the parabolic coordinate form (4.32). Thus, derivatives with respect to x, y are expressed by derivatives with respect to ξ, η. The formulas from section 2.1 are used to get these expressions:

$$\frac{\partial}{\partial x} = \vec{e}_x \cdot \vec{\nabla} = \frac{\eta \partial_\xi + \xi \partial_\eta}{\eta^2 + \xi^2}, \qquad \text{(A.106a)}$$
$$\frac{\partial}{\partial y} = \vec{e}_y \cdot \vec{\nabla} = \frac{\eta \partial_\eta - \xi \partial_\xi}{\eta^2 + \xi^2}. \qquad \text{(A.106b)}$$

We set $r_p^2 = \eta^2 + \xi^2$ and note $r_p^2 = 1 + x^2$ at the interface $\eta = \eta_s^{\text{Iv}} = 1$,

$$T^{l,\text{Iv}}(y_s) \approx \cancelto{=0}{T^{l,\text{Iv}}(y_s^{\text{Iv}})} + T_y^{l,\text{Iv}}(y_s^{\text{Iv}})\zeta = \left.\frac{\eta T_\eta^{l,\text{Iv}}}{r_p^2}\right|_{\eta=1} \zeta = -\frac{\zeta}{1+x^2}. \qquad \text{(A.107)}$$

We proceed:

$$T_y^{l,\text{Iv}}(y_s) \approx T_y^{l,\text{Iv}}(y_s^{\text{Iv}}) + T_{yy}^{l,\text{Iv}}(y_s^{\text{Iv}})\zeta. \qquad \text{(A.108)}$$

Appendix A. Auxiliary calculations

The first term in this expansion can be read from equation (A.107). The second term has to be calculated:

$$\begin{aligned}T_{yy}^{l,\text{Iv}}(y_s^{\text{Iv}}) &= \frac{1}{r_p^2}(\eta\partial_\eta - \xi\partial_\xi)\frac{\eta T_\eta^{l,\text{Iv}}}{r_p^2}\bigg|_{\eta=1} \\
&= \frac{1}{r_p^4}\left(\eta^2 T_{\eta\eta}^{l,\text{Iv}} + \eta T_\eta^{l,\text{Iv}}\right)_{\eta=1} \overset{\propto P_c}{\cancel{}} - \frac{2}{r_p^6}\left(\eta^2 - \xi^2\right)\eta T_\eta^{l,\text{Iv}}\bigg|_{\eta=1} \\
&= -\frac{1}{(1+x^2)^2} + 2\frac{1-x^2}{(1+x^2)^3} = \frac{1-3x^2}{(1+x^2)^3}\,.\end{aligned}$$
(A.109)

We insert into (A.108):

$$T_y^{l,\text{Iv}}(y_s) \approx -\frac{1}{1+x^2} + \frac{1-3x^2}{(1+x^2)^3}\zeta\,.$$
(A.110)

The last expression to calculate is $T_x^{l,\text{Iv}}(y_s)$:

$$T_x^{l,\text{Iv}}(y_s) \approx T_x^{l,\text{Iv}}(y_s^{\text{Iv}}) + T_{xy}^{l,\text{Iv}}(y_s^{\text{Iv}})\zeta\,.$$
(A.111)

The first term in (A.111) is

$$T_x^{l,\text{Iv}}(y_s^{\text{Iv}}) = \frac{\xi}{r_p^2}T_\eta^{l,\text{Iv}}\bigg|_{\eta=1} = -\frac{x}{1+x^2}$$
(A.112)

and the second term in (A.111) is

$$\begin{aligned}T_{xy}^{l,\text{Iv}}(y_s^{\text{Iv}}) &= \frac{1}{r_p^2}(\eta\partial_\eta - \xi\partial_\xi)\frac{\xi T_\eta^{l,\text{Iv}}}{r_p^2}\bigg|_{\eta=1} \\
&= \frac{1}{r_p^4}\left(\eta\xi T_{\eta\eta}^{l,\text{Iv}} - \xi T_\eta^{l,\text{Iv}}\right)_{\eta=1} \overset{\propto P_c}{\cancel{}} - \frac{2}{r_p^6}\left(\eta^2 - \xi^2\right)\xi T_\eta^{l,\text{Iv}}\bigg|_{\eta=1} \\
&= \frac{x}{(1+x^2)^2} + 2x\frac{1-x^2}{(1+x^2)^3} = x\frac{3-x^2}{(1+x^2)^3}\,.\end{aligned}$$
(A.113)

A.3. Thermal resistance in cartesian coordinates

We insert into (A.111):

$$T^{l,\mathrm{Iv}}_x(y_s) \approx -\frac{x}{1+x^2} + x\frac{3-x^2}{(1+x^2)^3}\zeta. \qquad (\mathrm{A.114})$$

Putting (A.107), (A.110) and (A.114) into the expanded versions of (4.28a)-(4.28c) (using $T \to T + T^{\mathrm{Iv}}$), we find

$$\sqrt{1+y_s'^2}\left(T^l - T^s - \frac{\zeta}{1+x^2}\right) \qquad \text{therm. dis-} \qquad (\mathrm{A.115a})$$
$$= \gamma_K(T_y^s - y_s'T_x^s) \qquad \text{continuity}$$
$$T^l = -\frac{1}{2}\sigma\kappa a(\theta) + \frac{\zeta}{1+x^2} \qquad \text{Gibbs-Thomson} \qquad (\mathrm{A.115b})$$
$$(\partial_y - y_s'\partial_x)\left(\mu_K T^s - T^l\right) = G_K\zeta + H_K\zeta' \qquad \text{asym. Stefan} \qquad (\mathrm{A.115c})$$

with

$$G_K = \frac{1-x^2}{(1+x^2)^2} \qquad (\mathrm{A.116a})$$

$$H_K = \frac{x}{1+x^2} - \frac{(3-x^2)x}{(1+x^2)^3}\zeta. \qquad (\mathrm{A.116b})$$

The dot products with $\vec{n}$ from (A.97) have already been calculated in these expressions.

The expanded bulk equations are simple:

$$T^{s,l}_{xx} + T^{s,l}_{yy} = 0. \qquad (\mathrm{A.117})$$

Here, P_c was set to zero. In the vicinity of the singularity at $x = -\mathrm{i}$, (A.117) can be replaced by its complex factorization (for details see section 2.2):

$$T^l_y = \mathrm{i}T^l_x, \qquad (\mathrm{A.118a})$$
$$T^s_y = -\mathrm{i}T^s_x. \qquad (\mathrm{A.118b})$$

Appendix A. Auxiliary calculations

We write down the total derivatives along the interface:

$$\frac{\mathrm{d}T^l}{\mathrm{d}x} = T^l_x + y'_s T^l_y = -\mathrm{i}(1 + \mathrm{i}y'_s)T^l_y, \tag{A.119a}$$

$$\frac{\mathrm{d}T^s}{\mathrm{d}x} = T^s_x + y'_s T^s_y = \mathrm{i}(1 - \mathrm{i}y'_s)T^s_y. \tag{A.119b}$$

(A.115c) is rewritten using (A.118a)-(A.118b):

$$-\mathrm{i}\mu_K \underbrace{(1 - \mathrm{i}y'_s)T^s_x}_{=\frac{\mathrm{d}T^s}{\mathrm{d}x}} - \mathrm{i}\underbrace{(1 + \mathrm{i}y'_s)T^l_x}_{=\frac{\mathrm{d}T^l}{\mathrm{d}x}} = G_K\zeta + H_K\zeta'. \tag{A.120}$$

This can be integrated directly using

$$\int \left(\frac{1 - x^2}{(1 + x^2)^2}\zeta + \frac{x}{1 + x^2}\zeta' \right) \mathrm{d}x = \frac{x}{1 + x^2}\zeta$$

and

$$\int \frac{(3 - x^2)x}{(1 + x^2)^3}\zeta\zeta' \,\mathrm{d}x = \frac{\zeta^2}{2}\frac{(3 - x^2)x}{(1 + x^2)^3}$$
$$- \frac{1}{2}\int \zeta^2 \frac{\mathrm{d}}{\mathrm{d}x}\left[\frac{(3 - x^2)x}{(1 + x^2)^3}\right] \mathrm{d}x = \mathcal{O}(\zeta^2) \to 0$$

and we find

$$\mu_K T^s + T^l = \mathrm{i}\zeta \frac{x}{1 + x^2} \tag{A.121}$$

or (solved for T^s)

$$T^s = \frac{1}{\mu_K}\left[\frac{1}{2}\sigma\kappa a(\theta) - \frac{(1 - \mathrm{i}x)\zeta}{1 + x^2}\right]. \tag{A.122}$$

In the last step, T^l was already replaced using (A.115b). The next step is to rewrite (A.115a) using (A.115b):

$$\sqrt{1 + y'^2_s}\left(-\frac{1}{2}\sigma\kappa a(\theta) - T^s\right) = \gamma_K\left[-\mathrm{i}(1 - \mathrm{i}y'_s)T^s_x\right]. \tag{A.123}$$

On the LHS of (A.123), T^s is replaced using (A.122) and on the RHS of (A.123) $-i(1 - iy'_s)T^s_x$ is replaced using (A.120):

$$\sqrt{1+y'^2_s}\left[-\frac{1}{2}\sigma\kappa a(\theta)\left(1+\frac{1}{\mu_K}\right) + \frac{(1-ix)\zeta}{\mu_K(1+x^2)}\right]$$
$$= \frac{\gamma_K}{\mu_K}\left[i(1+iy'_s)T^l_x + G_K\zeta + H_K\zeta'\right]. \quad (A.124)$$

Now, the expanded Gibbs-Thomson condition (A.115b) is differentiated along the interface:

$$T^l_x + y'_sT^l_y = (1+iy'_s)T^l_x = -\frac{1}{2}\sigma[\kappa a(\theta)]' + \frac{\zeta'}{1+x^2} - \frac{2x\zeta}{(1+x^2)^2}. \quad (A.125)$$

This is used to replace $(1+iy'_s)T^l_x$ in (A.124):

$$\sqrt{1+y'^2_s}\left[-\frac{1}{2}\sigma\kappa a(\theta)(1+\mu_K) + \frac{(1-ix)\zeta}{1+x^2}\right]$$
$$= \gamma_K\left[-\frac{i}{2}\sigma[\kappa a(\theta)]' + \frac{i\zeta'}{1+x^2} - \frac{2ix\zeta}{(1+x^2)^2}\right.$$
$$\left. + \frac{1-x^2}{(1+x^2)^2}\zeta + \frac{x}{1+x^2}\zeta' - \frac{(3-x^2)x}{(1+x^2)^3}\zeta\zeta'\right]$$
$$= i\gamma_K\left[-\frac{\sigma}{2}[\kappa a(\theta)]' + \left[\frac{\zeta}{1+x^2}\right]' - \left[\frac{ix\zeta}{1+x^2}\right]' + i\zeta\zeta'\frac{(3-x^2)x}{(1+x^2)^3}\right].$$

This is a less elegant form of (4.43).

A.4. Anisotropic diffusion

A.4.1. Expansion of the boundary conditions

We expand the boundary conditions of the corresponding symmetric model. The transition to the one-sided model can be made easily in the end. Let the superscripts n and s indicate the corresponding quantities

Appendix A. Auxiliary calculations

in the nematic phase and in the smectic-B phase, respectively. We need the partial derivatives of $T^{n,\mathrm{Iv}}(x,y)$ from (4.64). We use the abbreviation $r = \sqrt{x^2+y^2}$,

$$T^{n,\mathrm{Iv}}_x(x,y) = -\mathrm{e}^{\frac{P_\mu}{2}(1-y+r)} \frac{x}{2r\sqrt{y+r}},$$

$$T^{n,\mathrm{Iv}}_y(x,y) = -\mathrm{e}^{\frac{P_\mu}{2}(1-y+r)} \frac{\sqrt{y+r}}{2r}.$$

Now, these expressions have to be differentiated once more with respect to y:

$$T^{n,\mathrm{Iv}}_{xy}(x,y) = -T^{n,\mathrm{Iv}}_x(x,y)\left(P_\mu\frac{y+r}{2r} + \frac{r+2y}{2r^2}\right), \quad (\text{A.127a})$$

$$T^{n,\mathrm{Iv}}_{yy}(x,y) = -T^{n,\mathrm{Iv}}_y(x,y)\left(P_\mu\frac{y+r}{2r} - \frac{r-2y}{2r^2}\right). \quad (\text{A.127b})$$

Here, the second derivatives are reasonably expressed using the respective first derivative as a factor. Now, $y^{\mathrm{Iv}}_s(x) = \frac{1}{2}(1-x^2)$ is inserted. We use

$$\left.r\right|_{y^{\mathrm{Iv}}_s(x)} = \frac{1}{2}(1+x^2) \quad (\text{A.128a})$$

$$\left.\sqrt{y+r}\right|_{y^{\mathrm{Iv}}_s(x)} = 1 \quad (\text{A.128b})$$

and calculate the corresponding values:

$$T^{n,\mathrm{Iv}}_x\left(x, y^{\mathrm{Iv}}_s(x)\right) = -\frac{x}{1+x^2} \quad (\text{A.129a})$$

$$T^{n,\mathrm{Iv}}_y\left(x, y^{\mathrm{Iv}}_s(x)\right) = -\frac{1}{1+x^2} \quad (\text{A.129b})$$

$$T^{n,\mathrm{Iv}}_{xy}\left(x, y^{\mathrm{Iv}}_s(x)\right) = \frac{x}{(1+x^2)^2}\left(P_\mu + \frac{3-x^2}{1+x^2}\right) \quad (\text{A.130a})$$

$$T^{n,\mathrm{Iv}}_{yy}\left(x, y^{\mathrm{Iv}}_s(x)\right) = \frac{1}{(1+x^2)^2}\left(P_\mu + \frac{1-3x^2}{1+x^2}\right). \quad (\text{A.130b})$$

This can be used to expand the temperature in the nematic:

$$T^{n,\text{lv}}(x, y_s(x)) \approx T^{n,\text{lv}}(x, y_s^{\text{lv}}(x)) + T_y^{n,\text{lv}}((x, y_s^{\text{lv}}(x))\zeta(x) = -\frac{\zeta(x)}{1+x^2}. \tag{A.131}$$

Inserting this into $T^s = T^n$, and remembering $T \to T + T^{\text{lv}}$, directly yields the expanded continuity condition

$$T^s = T^n - \frac{\zeta(x)}{1+x^2}. \tag{A.132}$$

For the Stefan condition $1 = (\partial_y - y_s'(x)\partial_x)(T^s - T^n)$, we find

$$\begin{aligned}
1 &= (\partial_y - y_s'(x)\partial_x)(T^s - T^n) - T_y^{n,\text{lv}}(x, y_s^{\text{lv}}(x)) - T_{yy}^{n,\text{lv}}(x, y_s^{\text{lv}}(x))\zeta(x) \\
&\quad + (\zeta'(x) - x)\left(T_x^{n,\text{lv}}(x, y_s^{\text{lv}}(x)) + T_{xy}^{n,\text{lv}}(x, y_s^{\text{lv}}(x))\zeta(x)\right) \\
&= (\partial_y - y_s'(x)\partial_x)(T^s - T^n) + \frac{x}{1+x^2} - \frac{\zeta(x)}{(1+x^2)^2}\left(P_\mu + \frac{1-3x^2}{1+x^2}\right) \\
&\quad + (\zeta'(x) - x)\left[-\frac{x}{1+x^2} + \frac{x\zeta(x)}{(1+x^2)^2}\left(P_\mu + \frac{3-x^2}{1+x^2}\right)\right] \\
&= (\partial_y - y_s'(x)\partial_x)(T^s - T^n) + 1 - \frac{x\zeta'(x)}{1+x^2} - \frac{(1-x^2)\zeta(x)}{(1+x^2)^2} + \mathcal{O}(\zeta\zeta', P_\mu).
\end{aligned} \tag{A.133}$$

Using

$$\frac{x\zeta'(x)}{1+x^2} + \frac{(1-x^2)\zeta(x)}{(1+x^2)^2} = \left[\frac{x\zeta(x)}{1+x^2}\right]', \tag{A.134}$$

(A.133) becomes

$$\begin{aligned}
\left[\frac{x\zeta(x)}{1+x^2}\right]' &= (\partial_y - y_s'(x)\partial_x)(T^s - T^n) \\
&\quad + \frac{x(3-x^2)\zeta(x)\zeta'(x)}{(1+x^2)^3} - \frac{P_\mu \zeta(x)}{1+x^2}\left(1 - \frac{x\zeta'(x)}{1+x^2}\right).
\end{aligned} \tag{A.135}$$

This is the most general form of the expanded Stefan condition in the symmetric model, including nonlinear terms and terms of P_μ. The expanded Stefan condition (4.67b) is the linearized version of (A.135) in

Appendix A. Auxiliary calculations

the limit $P_\mu \to 0$ and with all T^s-terms dropped due to the one-sided model used in section 4.3. (A.135) is also valid in the case of isotropic diffusion, if P_μ is replaced by P_c ($\mu = 1$). Note, that these calculations are also carried out in appendix A.3.2 for the thermal resistance problem using more convenient but less complete expressions.

A.4.2. WKB solution

The curvature (4.61) in the homogeneous version of equation (4.73) is linearized:
$$\sigma_\mu \zeta''(x) + \frac{2(1+\mu^2 x^2)^{3/2}}{\mu(1+ix)} \zeta(x) = 0. \tag{A.136}$$

With $\sqrt{\sigma_\mu} = \varepsilon$, this results in the WKB form
$$S_0'^2 + \frac{2(1+\mu^2 x^2)^{3/2}}{\mu(1+ix)} + \varepsilon\left[S_0'' + 2S_0' S_1'\right] + \varepsilon^2 \left[S_1'' + S_1'^2\right] = 0. \tag{A.137}$$

This is solved iteratively order by order:

$$\varepsilon^0: \quad S_0'(x) = \left[-\frac{2(1+\mu^2 x^2)^{3/2}}{\mu(1+ix)}\right]^{\frac{1}{2}},$$
$$S_0(x) = i\int_{-i}^{x} \frac{\sqrt{2}(1+\mu^2 x'^2)^{3/4}}{\sqrt{\mu(1+ix')}} dx', \tag{A.138a}$$

$$\varepsilon^1: \quad S_1'(x) = -\frac{1}{2}\frac{d}{dx}\left[\ln\left(S_0'(x)\right)\right],$$
$$S_1(x) = \ln\left[\left(-\frac{\mu(1+ix)}{2(1+\mu^2 x^2)^{3/2}}\right)^{1/4}\right]. \tag{A.138b}$$

Using
$$\zeta(x) \sim e^{\frac{S_0(x)}{\sqrt{\sigma_\mu}} + S_1(x)} \qquad |x+i| \to \infty$$
one ends up with the WKB solution (4.74).

214

A.4.3. Stretching transformation of the shape equation

The transformation (4.75a)-(4.75b) is applied to equation (4.73). We restrict the expressions to leading terms in order to keep things more compact than in subsection A.2.4. First, we need the derivatives of $\zeta(x)$:

$$\zeta'(x) = -i\mu\bar{\sigma}^{\alpha_\mu}\dot{\phi}, \tag{A.139a}$$
$$\zeta''(x) = -\mu^2\ddot{\phi}. \tag{A.139b}$$

An important expression in the curvature and in the anisotropy functions is

$$\begin{aligned}(\zeta' - x)^2 &= \left(-i\mu\bar{\sigma}^{\alpha_\mu}\dot{\phi} - \frac{i}{\mu}(1 - \bar{\sigma}^{\alpha_\mu}t)\right)^2 \\ &= -\mu^2\bar{\sigma}^{2\alpha_\mu}\dot{\phi}^2 + 2\bar{\sigma}^{\alpha_\mu}\dot{\phi}(1 - \bar{\sigma}^{\alpha_\mu}t) - \frac{1}{\mu^2}\left(1 - 2\bar{\sigma}^{\alpha_\mu}t + \bar{\sigma}^{2\alpha_\mu}t^2\right) \\ &\approx -\frac{1}{\mu^2} + 2\bar{\sigma}^{\alpha_\mu}\left(\dot{\phi} + \frac{t}{\mu^2}\right).\end{aligned} \tag{A.140}$$

With these equations, one can easily get to the forms (4.76) and (4.77a)-(4.77b) for the curvature and the anisotropy functions respectively. For the right hand side, we need

$$1 + ix = 1 + \frac{1}{\mu}(1 - \bar{\sigma}^{\alpha_\mu}t) \approx \frac{1+\mu}{\mu} \tag{A.141}$$

and find

$$\frac{\zeta(x)}{1+ix} = \frac{\mu\bar{\sigma}^{2\alpha_\mu}\phi}{1+\mu} \tag{A.142}$$

for the RHS.

Appendix B.

C-code demo

The program for calculation of the numerical data presented in this work was written in C and compiled using the open source *GNU C compiler* (gcc). The numerical integration scheme was explained in detail in section 2.3. The ODE solver ``solvde'' as well as its driver routine ``relax'' were taken from *"Numerical recipes in C"* [96] and modified subsequently. Any memory allocation routines, complex number routines, algebraic routines and the false-position root finder were also taken from this book. However, user-provided functions representing the local system of differential equations are required. This section of the code is partly exhibited in this appendix chapter. The routines are shown exemplarily for the Oseen flow problem solved in the limit of small growth Péclet number P_c in section 3.2. They must be provided separately for integration on the line parallel to the imaginary axis and on the real axis. The corresponding code for the potential flow case (and also for anisotropic diffusion and kinetic effects numerics) was constructed using the same scheme. Comments in the files are given in German.

For the convective systems, the program has been developed further into an integrated tool, including a command line interface, a detailed and extensive documentation as well as an installation script for the program on *Linux*-based platforms. For further information, contact the author of this work, who created the program.

Appendix B. C-code demo

B.1. Oseen flow subroutines parallel to the imaginary axis

Local ODE system

```
#include <math.h>
#include <stdio.h>
#include "nrutil.h"
#include "complex_mod01.h"

extern float h1,        // Schrittweite
             x0,        // Kreuzungspunkt
             *x1,       // x-Achse
             F1R,F3R,   // Ableitungen
             b,         // Eigenwert
             P1,        // Peclet-Zahl
             **gsav;    // Integral
extern int M; // Anzahl Gitterpunkte

void difeq1_Oseen(int k, int k1, int k2, int jsf, int is1
   , int isf, int indexv[], int ne, float **s, float **y)
{
        fcomplex t,c1,
                 x_1,x_2,x_3,
                 f,f1,f2,f3,
                 g,g1,g2,u,v,d,
                 p1,p2,p3,p4,
                 E1,E2;

        c1.r=1.0;
        c1.i=0.0;

        t.r=x0;
        if(k>k2) t.i=x1[k-1];
        else t.i=x1[k];

        if(k==1) {
                x_1.r=y[1][1];
```

B.1. Oseen flow subroutines parallel to the imaginary axis

```
              x_1.i=y[2][1];
              x_2.r=y[3][1];
              x_2.i=y[4][1];
              x_3.r=y[5][1];
              x_3.i=y[6][1];
}
else if(k>k2) {
              x_1.r=y[1][k-1];
              x_1.i=y[2][k-1];
              x_2.r=y[3][k-1];
              x_2.i=y[4][k-1];
              x_3.r=y[5][k-1];
              x_3.i=y[6][k-1];
}
else {
              x_1.r=0.5*(y[1][k]+y[1][k-1]);
              x_1.i=0.5*(y[2][k]+y[2][k-1]);
              x_2.r=0.5*(y[3][k]+y[3][k-1]);
              x_2.i=0.5*(y[4][k]+y[4][k-1]);
              x_3.r=0.5*(y[5][k]+y[5][k-1]);
              x_3.i=0.5*(y[6][k]+y[6][k-1]);
}

p1=Cadd(Cmul(t,x_1),RCmul(P1,x_3));
p2=Cadd(t,x_1);
p3=Csub(c1,x_2);
p4=Cadd(c1,x_2);

u=RCmul(sqrt(2.0),Cmul(p1,Cmul(Cmul(Cmul(p4,p4),
   Cmul(p4,Csqrt(p4))),Cmul(Cmul(p3,Csqrt(p3)),
   Cmul(Cmul(p2,p2),Csqrt(p2))))));
v=Csub(Cmul(Cmul(p2,p2),Cmul(p4,p4)),RCmul(2.0*b,
   Cmul(p3,p3)));
d=Cdiv(Cmul(p3,Cmul(p4,p4)),RCmul(2.0,p2));

f=Csub(Cdiv(u,v),d);
f1=Cadd(Cdiv(d,p2),Cdiv(Cmul(u,Cadd(Cdiv(t,p1),
   Csub(Cdiv(RCmul(2.5,c1),p2),Cdiv(Cmul(Cmul(p4,
   p4),RCmul(2.0,p2)),v)))),v));
```

Appendix B. C-code demo

```
      f2=Cadd(Cmul(d,Csub(Cdiv(c1,p3),Cdiv(RCmul(2.0,c1
         ),p4))),Cdiv(Cmul(u,Csub(Cdiv(RCmul(3.5,c1),p4
         ),Cadd(Cdiv(RCmul(1.5,c1),p3),Cdiv
         (RCmul(2.0,Cadd(Cmul(p4,Cmul(p2,p2)),RCmul(2.0*b,
         p3))),v)))),v));
      f3=Cdiv(RCmul(P1,u),Cmul(v,p1));

      if(k<=(M+1)/2) integralg1_Oseen(&g,&g1,&g2,k,y);
      /* berechnet g's mit Trapezquadratur */
      else {
            g.r=*(*(gsav+1)+k);
            g.i=*(*(gsav+2)+k);
            g1.r=*(*(gsav+3)+k);
            g1.i=*(*(gsav+4)+k);
            g2.r=*(*(gsav+5)+k);
            g2.i=*(*(gsav+6)+k);
      }

      E1=Csub(p1,Cdiv(c1,RCmul(pow(2.0,1.5),Choch32(t))
         ));  /* komb. RB */
      E2=Cadd(Cadd(Cadd(x_1,Cmul(x_2,t)),RCmul(P1,g)),
         Cdiv(RCmul(3.0*pow(2.0,-2.5),c1),Cmul(t,Cmul(t
         ,Csqrt(t))))); /* komb. RB */

      if(k==k1) { /* Randbedingung am Anfang */
            s[4][6+indexv[1]]=t.r;
            s[4][6+indexv[2]]=-t.i;
            s[4][6+indexv[3]]=0.0;
            s[4][6+indexv[4]]=0.0;
            s[4][6+indexv[5]]=P1;
            s[4][6+indexv[6]]=0.0;
            s[4][jsf]=E1.r;

            s[5][6+indexv[1]]=t.i;
            s[5][6+indexv[2]]=t.r;
            s[5][6+indexv[3]]=0.0;
            s[5][6+indexv[4]]=0.0;
            s[5][6+indexv[5]]=0.0;
            s[5][6+indexv[6]]=P1;
```

B.1. Oseen flow subroutines parallel to the imaginary axis

```
            s[5][jsf]=E1.i;

            s[6][6+indexv[1]]=P1*g1.i;
            s[6][6+indexv[2]]=1.0+P1*g1.r;
            s[6][6+indexv[3]]=t.i+P1*g2.i;
            s[6][6+indexv[4]]=t.r+P1*g2.r;
            s[6][6+indexv[5]]=0.0;
            s[6][6+indexv[6]]=0.0;
            s[6][jsf]=E2.i;
    }
    else if(k>k2) { /* Randbedingung am Ende */
            s[1][6+indexv[1]]=0.0;
            s[1][6+indexv[2]]=1.0;
            s[1][6+indexv[3]]=0.0;
            s[1][6+indexv[4]]=0.0;
            s[1][6+indexv[5]]=0.0;
            s[1][6+indexv[6]]=0.0;
            s[1][jsf]=y[2][k-1]+y[2][1];

            s[2][6+indexv[1]]=0.0;
            s[2][6+indexv[2]]=0.0;
            s[2][6+indexv[3]]=0.0;
            s[2][6+indexv[4]]=1.0;
            s[2][6+indexv[5]]=0.0;
            s[2][6+indexv[6]]=0.0;
            s[2][jsf]=y[4][k-1]+y[4][1];

            s[3][6+indexv[1]]=0.0;
            s[3][6+indexv[2]]=0.0;
            s[3][6+indexv[3]]=0.0;
            s[3][6+indexv[4]]=0.0;
            s[3][6+indexv[5]]=0.0;
            s[3][6+indexv[6]]=1.0;
            s[3][jsf]=y[6][k-1]+y[6][1];
    }
    else { /* Matrix dazwischen */
            s[1][indexv[1]]=-1.0;
            s[1][indexv[2]]=0.0;
            s[1][indexv[3]]=0.0;
```

Appendix B. C-code demo

```
s[1][indexv[4]]=h1/2.0;
s[1][indexv[5]]=0.0;
s[1][indexv[6]]=0.0;
s[1][6+indexv[1]]=1.0;
s[1][6+indexv[2]]=0.0;
s[1][6+indexv[3]]=0.0;
s[1][6+indexv[4]]=h1/2.0;
s[1][6+indexv[5]]=0.0;
s[1][6+indexv[6]]=0.0;

s[2][indexv[1]]=0.0;
s[2][indexv[2]]=-1.0;
s[2][indexv[3]]=-h1/2.0;
s[2][indexv[4]]=0.0;
s[2][indexv[5]]=0.0;
s[2][indexv[6]]=0.0;
s[2][6+indexv[1]]=0.0;
s[2][6+indexv[2]]=1.0;
s[2][6+indexv[3]]=-h1/2.0;
s[2][6+indexv[4]]=0.0;
s[2][6+indexv[5]]=0.0;
s[2][6+indexv[6]]=0.0;

s[3][indexv[1]]=h1/2.0*f1.i;
s[3][indexv[2]]=h1/2.0*f1.r;
s[3][indexv[3]]=-1.0+h1/2.0*f2.i;
s[3][indexv[4]]=h1/2.0*f2.r;
s[3][indexv[5]]=h1/2.0*f3.i;
s[3][indexv[6]]=h1/2.0*f3.r;
s[3][6+indexv[1]]=s[3][indexv[1]];
s[3][6+indexv[2]]=s[3][indexv[2]];
s[3][6+indexv[3]]=2.0+s[3][indexv[3]];
s[3][6+indexv[4]]=s[3][indexv[4]];
s[3][6+indexv[5]]=s[3][indexv[5]];
s[3][6+indexv[6]]=s[3][indexv[6]];

s[4][indexv[1]]=-s[3][indexv[2]];
s[4][indexv[2]]=s[3][indexv[1]];
s[4][indexv[3]]=-s[3][indexv[4]];
```

B.1. Oseen flow subroutines parallel to the imaginary axis

```
s[4][indexv[4]]=s[3][indexv[3]];
s[4][indexv[5]]=-s[3][indexv[6]];
s[4][indexv[6]]=s[3][indexv[5]];
s[4][6+indexv[1]]=s[4][indexv[1]];
s[4][6+indexv[2]]=s[4][indexv[2]];
s[4][6+indexv[3]]=s[4][indexv[3]];
s[4][6+indexv[4]]=s[3][6+indexv[3]];
s[4][6+indexv[5]]=s[4][indexv[5]];
s[4][6+indexv[6]]=s[4][indexv[6]];

s[5][indexv[1]]=h1/2.0*g1.i;
s[5][indexv[2]]=h1/2.0*g1.r;
s[5][indexv[3]]=h1/2.0*g2.i;
s[5][indexv[4]]=h1/2.0*g2.r;
s[5][indexv[5]]=-1.0;
s[5][indexv[6]]=0.0;
s[5][6+indexv[1]]=s[5][indexv[1]];
s[5][6+indexv[2]]=s[5][indexv[2]];
s[5][6+indexv[3]]=s[5][indexv[3]];
s[5][6+indexv[4]]=s[5][indexv[4]];
s[5][6+indexv[5]]=2.0+s[5][indexv[5]];
s[5][6+indexv[6]]=s[5][indexv[6]];

s[6][indexv[1]]=-s[5][indexv[2]];
s[6][indexv[2]]=s[5][indexv[1]];
s[6][indexv[3]]=-s[5][indexv[4]];
s[6][indexv[4]]=s[5][indexv[3]];
s[6][indexv[5]]=-s[5][indexv[6]];
s[6][indexv[6]]=s[5][indexv[5]];
s[6][6+indexv[1]]=s[6][indexv[1]];
s[6][6+indexv[2]]=s[6][indexv[2]];
s[6][6+indexv[3]]=s[6][indexv[3]];
s[6][6+indexv[4]]=s[6][indexv[4]];
s[6][6+indexv[5]]=s[6][indexv[5]];
s[6][6+indexv[6]]=s[5][6+indexv[5]];

s[1][jsf]=y[1][k]-y[1][k-1]+h1*x_2.i;
s[2][jsf]=y[2][k]-y[2][k-1]-h1*x_2.r;
s[3][jsf]=y[3][k]-y[3][k-1]+h1*f.i;
```

Appendix B. C-code demo

```
                s[4][jsf]=y[4][k]-y[4][k-1]-h1*f.r;
                s[5][jsf]=y[5][k]-y[5][k-1]+h1*g.i;
                s[6][jsf]=y[6][k]-y[6][k-1]-h1*g.r;
        }
}
```

Integral g_k

```
void integralg1_Oseen(fcomplex *g, fcomplex *g1, fcomplex
    *g2, int k, float **y)
{
        int i,kspiegel=M+1-k;
        fcomplex x_1t, x_1tstrich, x_2t, x_2tstrich, t,
            tstrich, l[3];
        x_1t=Complex(0.5*(y[1][kspiegel]+y[1][kspiegel
            -1]),0.5*(y[2][kspiegel]+y[2][kspiegel-1]));
        x_2t=Complex(0.5*(y[3][kspiegel]+y[3][kspiegel
            -1]),0.5*(y[4][kspiegel]+y[4][kspiegel-1]));
        t=Complex(x0,x1[kspiegel]);

        dljk1_Oseen(Complex(y[1][M],y[2][M]),x_2t,Complex
            (y[3][M],y[4][M]),t,Complex(x0,x1[M]),l); /*
            berechne Integrand am Endpunkt konsistent fuer
            alle k */
        *g=*(l+0);
        *g1=*(l+1);
        *g2=*(l+2);

        dlkk1_Oseen(x_1t,x_2t,l); /* berechne Integrand
            am Anfangspunkt */
        *g=Cadd(*g,*(l+0));
        *g1=Cadd(*g1,*(l+1));
        *g2=Cadd(*g2,*(l+2));

        *g=RCmul(0.5,*g);
        *g1=RCmul(0.5,*g1);
        *g2=RCmul(0.5,*g2);

        for(i=kspiegel+1;i<M;i++) { /* berechne Integrand
```

B.1. Oseen flow subroutines parallel to the imaginary axis

```
                  auf Zwischenpunkten */
                    x_1tstrich=Complex(0.5*(y[1][i]+y[1][i
                       -1]),0.5*(y[2][i]+y[2][i-1]));
                    x_2tstrich=Complex(0.5*(y[3][i]+y[3][i
                       -1]),0.5*(y[4][i]+y[4][i-1]));
                    tstrich=Complex(x0,x1[i]);
                    dljk1_Oseen(x_1tstrich,x_2t,x_2tstrich,t,
                       tstrich,1);
                    *g=Cadd(*g,*(l+0));
                    *g1=Cadd(*g1,*(l+1));
                    *g2=Cadd(*g2,*(l+2));
                }
//          if(k==kspiegel) { /* g muss nicht stetig bei Im(t
     )=0 sein! */
//                  g->i=0.0;
//                  g1->i=0.0;
//                  g2->i=0.0;
//          }

            g->r*=h1; /* Trapezregel zu Ende und stelle
               richtige Symmetrie her */
            g->i*=-h1;
            g1->r*=h1;
            g1->i*=-h1;
            g2->r*=h1;
            g2->i*=-h1;

            *(*(gsav+1)+kspiegel)=g->r; /* speichere Werte
               von g fuer obere Seite */
            *(*(gsav+2)+kspiegel)=-g->i;
            *(*(gsav+3)+kspiegel)=g1->r;
            *(*(gsav+4)+kspiegel)=-g1->i;
            *(*(gsav+5)+kspiegel)=g2->r;
            *(*(gsav+6)+kspiegel)=-g2->i;

            *(*(gsav+1)+k)=g->r; /* speichere Werte von g
               fuer untere Seite (Symmetrie) */
            *(*(gsav+2)+k)=g->i;
```

Appendix B. C-code demo

```
                *(*(gsav+3)+k)=g1->r;
                *(*(gsav+4)+k)=g1->i;
                *(*(gsav+5)+k)=g2->r;
                *(*(gsav+6)+k)=g2->i;
}

void dlkk1_Oseen(fcomplex x_1t, fcomplex x_2t, fcomplex l
    [])
{
        int i=0;
        fcomplex p3;

        p3=RCsub(1.0,x_2t);

        *(l+0)=Cmul(p3,x_1t);
        *(l+1)=p3;
        *(l+2)=RCmul(-1.0,x_1t);

        while(i<3) {
                *(l+i)=Cmul(*(l+i),Complex(0.0,-1.0));
                i++;
        }
}

void dljk1_Oseen(fcomplex x_1tstrich, fcomplex x_2t,
    fcomplex x_2tstrich, fcomplex t, fcomplex tstrich,
    fcomplex l[])
{
        int i=0;
        fcomplex p3,p4;

        p3=RCsub(1.0,x_2t);
        p4=RCadd(1.0,x_2tstrich);

        *(l+0)=Cmul(p3,Cadd(Cdiv(Cmul(x_2tstrich,Csub(t,
            tstrich)),p4),x_1tstrich));
        *(l+1)=Complex(0.0,0.0);
        *(l+2)=Cdiv(RCmul(-1.0,*(l+0)),p3);
```

```
          while(i<3) {
                  *(l+i)=Cmul(*(l+i),Complex(0.0,-1.0));
                  i++;
          }
}
```

B.2. Oseen flow subroutines on the real axis

Local ODE system

```
#include <math.h>
#include <stdio.h>
#include "nrutil.h"
#include "complex_mod01.h"

extern float h2,      // Schrittweite
             x0,      // Kreuzungspunkt
             *x2,     // x-Achse
             F1R,F3R, // Ableitungen
             b,       // Eigenwert
             P1,      // Peclet-Zahl
             **gsav_reell;  // Integral
extern int M; // Anzahl Gitterpunkte

void difeq2_Oseen(int k, int k1, int k2, int jsf, int is1
    , int isf, int indexv[], int ne, float **s, float **y)
{
        float t,
              x_1,x_2,x_3,
              f,f1,f2,f3,
              g,g1,g2,
              u,v,d,
              p1,p2,p3,p4,
              E1;

        t=x2[k];
        if(k>k2)  t=x2[k-1];

        if(k==1) {
```

Appendix B. C-code demo

```c
            x_1=y[1][1];
            x_2=y[2][1];
            x_3=y[3][1];
    }
    else if(k>k2) {
            x_1=y[1][k-1];
            x_2=y[2][k-1];
            x_3=y[3][k-1];
    }
    else {
            x_1=0.5*(y[1][k]+y[1][k-1]);
            x_2=0.5*(y[2][k]+y[2][k-1]);
            x_3=0.5*(y[3][k]+y[3][k-1]);
    }

    p1=x_1*t+P1*x_3;
    p2=t+x_1;
    p3=1.0-x_2;
    p4=1.0+x_2;

    u=sqrt(2.0)*p1*pow(p2,2.5)*pow(p3,1.5)*pow(p4
        ,3.5);
    v=pow(p2*p4,2.0)-2.0*b*pow(p3,2.0);
    d=(pow(p4,2.0)*p3)/(2.0*p2);

    f=u/v-d;
    f1=d/p2+(u/v)*(t/p1+2.5/p2-2.0*pow(p4,2.0)*p2/v);
    f2=d*(1.0/p3-2.0/p4)+(u/v)*(3.5/p4-1.5/p3-2.0*(p4
        *pow(p2,2.0)+2.0*b*p3)/v);
    f3=(u*P1)/(v*p1);

    if(k>1) integralg2_Oseen(&g,&g1,&g2,k,y);

    E1=p1-pow(2.0*t,-1.5);   /* komb. RB */

    if(k==k1) { /* Randbedingung am Anfang */
            s[2][ne+indexv[1]]=1.0;
            s[2][ne+indexv[2]]=0.0;
            s[2][ne+indexv[3]]=0.0;
```

B.2. Oseen flow subroutines on the real axis

```
        s[2][jsf]=y[1][1]-F1R;

        s[3][ne+indexv[1]]=0.0;
        s[3][ne+indexv[2]]=0.0;
        s[3][ne+indexv[3]]=1.0;
        s[3][jsf]=y[3][1]-F3R;
}

else if(k>k2) { /* Randbedingung am Ende */
        s[1][ne+indexv[1]]=t; /* komb. RB */
        s[1][ne+indexv[2]]=0.0;
        s[1][ne+indexv[3]]=P1;
        s[1][jsf]=E1;
}

else { /* Matrix dazwischen */
        s[1][indexv[1]]=-1.0;
        s[1][indexv[2]]=-h2/2.0;
        s[1][indexv[3]]=0.0;
        s[1][ne+indexv[1]]=1.0;
        s[1][ne+indexv[2]]=-h2/2.0;
        s[1][ne+indexv[3]]=0.0;

        s[2][indexv[1]]=-h2/2.0*f1;
        s[2][indexv[2]]=-1.0-h2/2.0*f2;
        s[2][indexv[3]]=-h2/2.0*f3;
        s[2][ne+indexv[1]]=s[2][indexv[1]];
        s[2][ne+indexv[2]]=2.0+s[2][indexv[2]];
        s[2][ne+indexv[3]]=s[2][indexv[3]];

        s[3][indexv[1]]=-h2/2.0*g1;
        s[3][indexv[2]]=-h2/2.0*g2;
        s[3][indexv[3]]=-1.0;
        s[3][ne+indexv[1]]=s[3][indexv[1]];
        s[3][ne+indexv[2]]=s[3][indexv[2]];
        s[3][ne+indexv[3]]=1.0;

        s[1][jsf]=y[1][k]-y[1][k-1]-h2*x_2;
        s[2][jsf]=y[2][k]-y[2][k-1]-h2*f;
```

Appendix B. C-code demo

```
                s[3][jsf]=y[3][k]-y[3][k-1]-h2*g;
        }
}
```

Integral g_k

```
void integralg2_Oseen(float *g, float *g1, float *g2, int
    k, float **y)
{
        int i;
        float x_1t, x_1tstrich, x_2t, x_2tstrich, t,
            tstrich, l[3];
        x_1t=0.5*(y[1][k]+y[1][k-1]);
        x_2t=0.5*(y[2][k]+y[2][k-1]);
        t=x2[k];

        dljk2_Oseen(y[1][M],x_2t,y[2][M],t,x2[M],l); /*
            berechne Integrand am Endpunkt konsistent fuer
            alle k */
        *g=*(l+0);
        *g1=*(l+1);
        *g2=*(l+2);

        dlkk2_Oseen(x_1t,x_2t,l); /* berechne Integrand
            am Anfangspunkt */
        *g+=*(l+0);
        *g1+=*(l+1);
        *g2+=*(l+2);

        *g*=0.5;
        *g1*=0.5;
        *g2*=0.5;

        for(i=k+1;i<M;i++) { /* berechne Integrand auf
            Zwischenpunkten */
                x_1tstrich=0.5*(y[1][i]+y[1][i-1]);
                x_2tstrich=0.5*(y[2][i]+y[2][i-1]);
                tstrich=x2[i];
                dljk2_Oseen(x_1tstrich,x_2t,x_2tstrich,t,
```

```
                        tstrich,1);
                *g+=*(l+0);
                *g1+=*(l+1);
                *g2+=*(l+2);
        }

        *g*=h2;
        *g1*=h2;
        *g2*=h2;

        *(*(gsav_reell+1)+k)=*g;  /* speichere Werte von g
           fuer obere Seite */
        *(*(gsav_reell+2)+k)=*g1;
        *(*(gsav_reell+3)+k)=*g2;
}

void dlkk2_Oseen(float x_1t, float x_2t, float l[])
{
        float p3;

        p3=1.0-x_2t;

        *(l+0)=-p3*x_1t;  // bedenke l=-lSchlange
        *(l+1)=-p3;
        *(l+2)=x_1t;
}

void dljk2_Oseen(float x_1tstrich, float x_2t, float
   x_2tstrich, float t, float tstrich, float l[])
{
        float p3,p4;

        p3=1.0-x_2t;
        p4=1.0+x_2tstrich;

        *(l+0)=-p3*(x_2tstrich*(t-tstrich)/p4+x_1tstrich)
           ;  // bedenke l=-lSchlange
        *(l+1)=0.0;
        *(l+2)=-*(l+0)/p3;
```

Appendix B. C-code demo

}

Appendix C.

Material parameters

In this appendix chapter, the material parameters of all the substances under consideration in this work are gathered in one big table (table C.1). The experiments, in which they were measured, are cited in the table, too.

d_0 capillary length
D thermal diffusivity (pure substances) or impurity diffusivity (solutions)
β surface tension anisotropy strength
ρ_m mass density
L/c ratio of latent heat and specific heat
T_M equilibrium melting temperature
γ_0 mean surface tension averaged over all interface normal orientations
Pr Prandtl number
ν kinematic viscosity
μ heat diffusion anisotropy strength

Appendix C. Material parameters

Table C.1.: Experimental values (with citations) of material parameters for various substances used as input for the numerical calculations in this work: pivalic acid (PVA), succinonitrile (SCN), ammonium bromide (NH_4Br), helium-3 (^{3}He) and the liquid crystal CCH5

	PVA	SCN	NH_4Br	^{3}He	CCH5
$d_0 [nm]$	3.76 [101] (pure) 3.2 [18] (with 2 % C_2H_5OH) 1.7 [18] (with 4 % C_2H_5OH)	2.79 [101] (pure) 13 [41] (ethylene mix)	0.28 [40] (51 % in H_2O)	3.8 [99]	$5 \cdot 10^{-3}$ [28]
$D [\mu m^2/s]$	$7 \cdot 10^4$ [101] (pure) 250 [18] (2-4 % C_2H_5OH, 20°C)	$1.13 \cdot 10^5$ [101]	2600 [40] (51 % in H_2O)	$5.46 \cdot 10^4$ [59] (at 0.1 K) $5.46 \cdot 10^4$ [59] (at 0.5 K)	$1.25 \cdot 10^5$ [100] (for K15)
β	0.09 [39] 0.375 [91] 0.75 [52]	0.075 [101]	0.24 [40] (51 % in H_2O)	0.3 [99]	$\beta_2 = 0.06 \ldots 0.18$ [17] $\beta_6 = 0.035$ [17]
$\rho_m [kg/m^3]$	908 [101]				1000
$L/c [K]$	11.83 [115]	23.12 [102]			11.26 [121]
$T_M [°C]$	35.935 [115]	58.1 [64]		-272.83 [99]	51.2 [121]
$\gamma_0 [J/m^2]$	$2.84 \cdot 10^{-3}$ [114]	$8.9 \cdot 10^{-3}$ [102]		$6 \cdot 10^{-5}$ [98]	$3 \cdot 10^{-6}$ [28]
Pr	134.92 [101]	23.03 [107]	343.5 ($\nu(H_2O$ at 25°C)/ $D(51$ % in H_2O))		
$\nu [\mu m^2/s]$	$3 \cdot 10^6$ [115]	$26 \cdot 10^5$ [120]	$8.931 \cdot 10^5$ [42] (H_2O at 25°C) $5.537 \cdot 10^5$ [42] (H_2O at 50°C)		
μ					0.767 [100] (for K15)

Bibliography

[1] T. Abel, E. Brener, and H. Müller-Krumbhaar. Three-dimensional growth morphologies in diffusion-controlled channel growth. *Phys. Rev. E*, 55(6):7789–7792, 1997.

[2] D. V. Alexandrov and P. K. Galenko. Selection criterion of stable dendritic growth at arbitrary Péclet numbers with convection. *Phys. Rev. E*, 87:062403, 2013.

[3] D.V. Alexandrov, P.K. Galenko, and D.M. Herlach. Selection criterion for the growing dendritic tip in a non-isothermal binary system under forced convective flow. *J. Crys. Growth*, 312:2122–2127, 2010.

[4] M. Ben Amar. Dendritic growth rate at arbitrary undercooling. *Phys. Rev. A*, 41(4):2080–2092, 1990.

[5] M. Ben Amar, Ph. Bouissou, and P. Pelcé. An exact solution for the shape of a crystal growing in a forced flow. *J. Crys. Growth*, 92:97–100, 1988.

[6] M. Ben Amar and B. Moussallam. Numerical results on two-dimensional dendritic solidification. *Physica*, 25D:155–164, 1987.

[7] M. Ben Amar and Y. Pomeau. Theory of dendritic growth in a weakly undercooled melt. *Europhys. Lett.*, 2(4):307–314, 1986.

Bibliography

[8] Martine Ben Amar and Efim Brener. Theory of pattern selection in three-dimensional nonaxissymmetric denritic growth. *Phys. Rev. Lett.*, 71(4):589–592, 1993.

[9] Martine Ben Amar and Efim Brener. Parity-broken dendrites. *Phys. Rev. Lett.*, 17(3):561–564, 1995.

[10] R. Ananth and W. N. Gill. Self-consistent theory of dendritic growth with convection. *J. Crys. Growth*, 108:173–189, 1991.

[11] M. Asta, C. Beckermann, A. Karma, W. Kurz, R. Napolitano, M. Plapp, G. Purdy, M. Rappaz, and R. Trivedi. Solidification microstructures and solid-state parallels: Recent developments, future directions. *Acta Materialia*, 57(146):941–971, 2009.

[12] S. Balibar, D. O. Edwards, and W. F. Saam. The effect of heat flow and nonhydrostatic strain on the surface of helium crystals. *J. Low Temp. Phys.*, 82(3–4):119–143, 1991.

[13] M. N. Barber, A. Barbieri, and J. S. Langer. Dynamics of dendritic sidebranching in the two-dimensional symmetric model of solidification. *Phys. Rev. A*, 36:3340, 1987.

[14] A. Barbieri and J. S. Langer. Predictions of dendritic growth rates in the linearized solvability theory. *Phys. Rev. A*, 39(10):5314–5325, 1989.

[15] Angelo Barbieri, Daniel C. Hong, and J. S. Langer. Velocity selection in the symmetric model of dendritic crystal growth. *Phys. Rev. A*, 35(4):1802–1808, 1987.

[16] J. H. Bilgram, M. Firmann, and E. Hürlimann. Dendritic solidification of Krypton and Xenon. *J. Crys. Growth*, 96:175–185, 1989.

[17] T. Börzsönyi, Á. Buka, and L. Kramer. Effect of the anisotropic surface tension, crystallization kinetics, and heat diffusion on nonequilibrium growth of liquid crystals. *Phys. Rev. E*, 58(5):6236–6245, 1998.

[18] P. Bouissou, B. Perrin, and P. Tabeling. Influence of an external flow on dendritic crystal growth. *Phys. Rev. A*, 40(1):509, 1989.

[19] Ph. Bouissou and P. Pelcé. Effect of a forced flow on dendritic growth. *Phys. Rev. A*, 40(11):6673–6680, 1989.

[20] E. Brener, G. Boussinot, C. Hüter, M. Fleck, D. Pilipenko, R. Spatschek, and D. E Temkin. Pattern formation during diffusional transformations in the presence of triple junctions and elastic effects. *J. Phys.: Condens. Matter*, 21:464106, 2009.

[21] E. Brener, H. Müller-Krumbhaar, and D. Temkin. Structure formation and the morphology diagram of possible structures in two-dimensional diffusional growth. *Phys. Rev. E*, 54(3):2714–2722, 1996.

[22] E. A. Brener. Effects of surface energy and kinetics on the growth of needle-like dendrites. *J. Crys. Growth*, 90:165–170, 1990.

[23] E. A. Brener, C. Hüter, D. Pilipenko, and D. E. Temkin. Velocity selection problem in the presence of the triple junction. *Phys. Rev. Lett.*, 99:105701, 2007.

[24] E. A. Brener and V. I. Mel'nikov. Pattern selection in two-dimensional dendritic growth. *Advances in Physics*, 40(1):53–97, 1991.

[25] Efim Brener. Needle-crystal solution in three-dimensional dendritic growth. *Phys. Rev. Lett.*, 71(22):3653–3656, 1993.

[26] Efim Brener and Herbert Levine. Growth of non-reflection symmetric dendrites. *Phys. Rev. A*, 43(2):883–887, 1991.

[27] Efim A. Brener and D. E. Temkin. Velocity-selection problem for combined motion of melting and solidification fronts. *Phys. Rev. Lett.*, 94:184501, 2005.

[28] Á. Buka, T. Tóth-Katona, and L. Kramer. Equilibrium shapes of a nematic-smectic-b liquid-crystal interface. *Phys. Rev. E*, 49(6):5271–5275, 1994.

[29] B. Caroli, C. Caroli, B. Roulet, and J. S. Langer. Solvability condition for needle crystals at large undercooling in a nonlocal model of solidification. *Phys. Rev. A*, 33(1):442–452, 1986.

[30] B. Castaing and P. Nozières. Transmission of sound at the liquid-solid interface of helium : a new probe of melting kinetics. *J. Phys. France*, 41(7):701–706, 1980. DOI: 10.1051/jphys:01980004107070100.

[31] S. J. Chapman. Asymptotic analysis of the ginzburg-landau model of superconductivity: reduction to a free-boundary model. *Quart. Appl. Math.*, LIII:601–627, 1995.

[32] J. Cognard. The anisotropy of the surface tension of polar liquids: The case of liquid crystals. *J. Adhesion*, 17:123–134, 1984.

[33] R. Combescot, V. Hakim, T. Dombre, Y. Pomeau, and A. Pumir. Analytic theory of the Saffman-Taylor fingers. *Phys. Rev. A*, 37(4):1270–1283, 1988.

[34] Roland Combescot, Thierry Dombre, Vincent Hakim, Yves Pomeau, and Alain Pumir. Shape selection of Saffman-Taylor fingers. *Phys. Rev. Lett.*, 56(19):2036–2039, 1986.

[35] S. R. Coriell and D. Turnbull. Relative roles of heat transport and interface rearrangement rates in the rapid growth of crystals in undercooled melts. *Acta Metall.*, 30:2135–2139, 1982.

[36] Jean-Marc Debierre, Rahma Guérin, and Klaus Kassner. Crystal growth in a channel: Pulsating fingers, merry-go-round patterns, and seesaw dynamics. *Phys. Rev. E*, 88:042407, 2013. DOI: 10.1103/PhysRevE.88.042407.

[37] T. Dombre, V. Hakim, and Y. Pomeau. Séléction de la largeur des doigts dans l'instabilité de Saffman-Taylor. *C.R. Acad. Sci.*, 302:803–808, 1986.

[38] Alan T. Dorsey and Olivier Martin. Saffman-Taylor fingers with anisotropic surface tension. *Phys. Rev. A*, 35(9):3989–3992, 1987.

[39] A. Dougherty. Surface tension anisotropy and the dendritic growth of pivalic acid. *J. Crys. Growth*, 110:501–508, 1991.

[40] A. Dougherty and J. P. Gollub. Steady-state dendritic growth of NH_4Br from solution. *Phys. Rev. A*, 38(6):3043–3053, 1988.

[41] Blas Echebarria, Roger Folch, Alain Karma, and Mathis Plapp. Quantitative phase-field model of alloy solidification. *Phys. Rev. E*, 70:061604, 2004.

[42] Virginie Emsellem and Patrick Tabeling. Experimental study of dendritic growth with an external flow. *J. Crys. Growth*, 156:286–295, 1995.

[43] T. Fischaleck. *An approach to selection theory for dendritic growth enabling the treatment of general bulk equations*. PhD thesis, Otto-von-Guericke-University Magdeburg, Faculty of

natural Sciences, Intitute for theoretical Physics, computational Physics group, Germany, 2008.

[44] Thomas Fischaleck and Klaus Kassner. Extending the scope of microscopic solvability: Combination of the Kruskal-Segur method with Zauderer decomposition. *Europhys. Lett.*, 81:54004, 2008.

[45] O. Funke, G. Phanikumar, P.K. Galenko, L. Chernova, S. Reutzel, M. Kolbe, and D.M. Herlach. Dendrite growth velocity in levitated undercooled nickel melts. *J. Crys. Growth*, 297:211–222, 2006.

[46] P. K. Galenko, O. Funke, J. Wang, and D. M. Herlach. Kinetics of dendritic growth under the influence of convective flow in solidification of undercooled droplets. *Materials Science and Engineering A*, 375–377:488–492, 2004.

[47] M. Georgelin and A. Pocheau. Onset of sidebranching in directional solidification. *Phys. Rev. E*, 57(3):3189–3203, 1998.

[48] N. Gheorghiu and J. T. Gleeson. Length and speed selection in dendritic growth of electrohydrodynamic convection in a nematic liquid crystal. *Phys. Rev. E*, 66:051710, 2002.

[49] R. Glardon and W. Kurz. Solidification path and phase diagram of directionally solidified Co-Sm-Cu alloys. *J. Crys. Growth*, 51:283–291, 1981.

[50] M. E. Glicksman, M. B. Koss, L. T. Bushnell, J. C. Lacombe, and E. A. Winsa. Dendritic growth of succinonitrile in terrestrial and microgravity conditions as a test of theory. *ISIJ Int.*, 35(6):604–610, 1995.

[51] M. E. Glicksman, R. J. Schaefer, and J. D. Ayers. Dendritic growth - a test of theory. *Metall. Trans. A*, 7(11):1747, 1976.

[52] M. E. Glicksman and N. B. Singh. Effects of crystal-melt interfacial energy anisotropy on dendritic morphology and growth kinetics. *J. Crys. Growth*, 98:277–284, 1989.

[53] Martin Eden Glicksman. *Principles of Solidification*. Springer Science + Business Media, New York, 2011.

[54] O. A. Gomes, R. C. Falco, and O. N. Mesquita. Anomalous capillary length in cellular nematic-isotropic interfaces. *Phys. Rev. Lett.*, 86(12):2577–2580, 2001.

[55] R. González-Cinca, L. Ramírez-Piscina, J. Casademunt, A. Hernández-Machado, and Á. Buka T. Tóth-Katona, T. Börzsönyi. Heat diffusion anisotropy in dendritic growth: phase field simulations and experiments in liquid crystals. *J. Crys. Growth*, 193:712–719, 1998.

[56] F. Graner, S. Balibar, and E. Rolley. The growth rate of ^{3}He crystals. *J. Low Temp. Phys.*, 75(1/2):69–77, 1989.

[57] F. Graner, R. M. Bowley, and P. Nozières. The growth kinetics of ^{3}He crystals. *J. Low Temp. Phys.*, 80(3–4):113, 1990.

[58] Dennis S. Greywall. Specific heat of normal liquid ^{3}He. *Phys. Rev. B*, 27(5):2747–2766, 1983.

[59] Dennis S. Greywall. Thermal conductivity of normal liquid ^{3}He. *Phys. Rev. B*, 29(9):4933–4945, 1984.

[60] Sebastian Gurevich, Alain Karma, Mathis Plapp, and Rohit Trivedi. Phase-field study of three-dimensional steady-state growth shapes in directional solidification. *Phys. Rev. E*, 81:011603, 2010. DOI: 10.1103/PhysRevE.81.011603.

Bibliography

[61] D. C. Hong and J. S. Langer. Analytic theory of the selection mechanism in the Saffman-Taylor problem. *Phys. Rev. Lett.*, 56(19):2032–2035, 1986.

[62] D. C. Hong and J. S. Langer. Pattern selection and tip perturbations in the Saffman-Taylor problem. *Phys. Rev. A*, 36:2325, 1987.

[63] Harudo Honjo and Yasuji Sawada. Quantitative measurements on the morphology of a NH_4Br dendritic crystal growth in a capillary. *J. Crys. Growth*, 58:297–303, 1982.

[64] S. C. Huang and M. E. Glicksman. Fundamentals of dendritic solidification. *Acta Metall.*, 29:701–717, 1981.

[65] John B. Hudson. *Surface Science*. Reed Publishing Inc., USA, 1992.

[66] G. P. Ivantsov. Temperature field around a spheroidal, cylindrical and acicular crystal growing in a supercooled melt. *Dokl. Akad. Nauk SSSR*, 58:567, 1947.

[67] Alain Karma and Wouter-Jan Rappel. Phase-field method for computationally efficient modeling of solidification with arbitrary interface kinetics. *Phys. Rev. E*, 53(4):3017–3020, 1996.

[68] Alain Karma and Wouter-Jan Rappel. Quantitative phase-field modeling of dendritic growth in two and three dimensions. *Phys. Rev. E*, 57(4):4323–4349, 1998.

[69] Klaus Kassner. *Pattern Formation in diffusion-limited Crystal Growth*. World Scientific, Singapore, 1996.

[70] Klaus Kassner, Rahma Guérin, Tristan Ducousso, and Jean-Marc Debierre. Phase-field study of solidification in three-

dimensional channels. *Phys. Rev. E*, 82:021606, 2010. DOI: 10.1103/PhysRevE.82.021606.

[71] David A. Kessler and Herbert Levine. Stability of dendritic crystals. *Phys. Rev. Lett.*, 57(24):3069–3072, 1986.

[72] David A. Kessler and Herbert Levine. Velocity selection in dendritic growth. *Phys. Rev. B*, 33(11):7867–7870, 1986.

[73] M. B. Koss, L. T. Bushnell, J. C. Lacombe, and M. E. Glicksman. The effect of convection on dendritic growth under microgravity conditions. *Chem. Eng. Comm.*, 152–153(1):351–363, 1996.

[74] M. D. Kruskal and H. Segur. Asymptotics beyond all orders. *Physica*, 28D:228, 1987.

[75] M. D. Kruskal and H. Segur. Asymptotics beyond all orders in a model of crystal growth. *Stud. Appl. Math.*, 85:129, 1991.

[76] Douglas A. Kurtze. Needle crystals with nonlinear diffusion. *Phys. Rev. A*, 36(1):232–237, 1987.

[77] W. Kurz and D. J. Fischer. *Fundamentals of Solidification*. Trans Tech Publications Ltd., Switzerland, 3. edition, 1992.

[78] J. C. LaCombe, M. B. Koss, J. E. Frei, C. Giummarra, A. O. Lupulescu, and M. E. Glicksman. Evidence for tip velocity oscillations in dendritic solidification. *Phys. Rev. E*, 65:031604, 2002.

[79] H. Lamb. *Hydrodynamics*. Cambridge Univ. Press, London, 6. edition, 1932.

Bibliography

[80] C. W. Lan, C. M. Hsu, and C. C. Liu. Efficient adaptive phase field simulation of dendritic growth in a forced flow at low supercooling. *J. Crys. Growth*, 241:379–386, 2002.

[81] J. S. Langer. Instabilities and pattern formation in crystal growth. *Rev. Mod. Phys.*, 52(1):1–28, 1980a.

[82] J. S. Langer. Lectures in the theory of pattern formation. In J. Souletie, J. Vannimenus, and R. Stora, editors, *Les Houches - Le hasard et la matière/Chance and matter*, Session XLVI 1986, pages 628–711. Elsevier Science Publishers B. V., 1987. Course 10.

[83] J. S. Langer and H. Müller-Krumbhaar. Theory of dendritic growth. *Acta Metall.*, 26:1681,1689,1697, 1978.

[84] Youn-Woo Lee, Ramagopal Ananth, and William N. Gill. Selection of a length scale in unconstrained dendritic growth with convection in the melt. *J. Crys. Growth*, 123:226–230, 1993.

[85] Q. Li and C. Beckermann. Modeling of free dendritic growth of succinonitrile-acetone alloys with thermosolutal melt convection. *J. Crys. Growth*, 236:482–498, 2002.

[86] Dmitry Medvedev, Thomas Fischaleck, and Klaus Kassner. Influence of external flows on crystal growth: Numerical investigation. *Phys. Rev. E*, 74:031606, 2006.

[87] Daniel I. Meiron. Boundary integral formulation of the two-dimensional symmetric model of dendritic growth. *Physica*, 23D:329–339, 1986.

[88] C. Misbah. Velocity selection for needle crystals in the 2-D one-sided model. *J. Physique*, 48:1265–1272, 1987.

[89] P. Molho, A. J. Simon, and A. Libchaber. Péclet number and crystal growth in a channel. *Phys. Rev. A*, 42(2):904–910, 1990.

[90] W. W. Mullins and R. F. Sekerka. Stability of a planar interface during solidification of a dilute binary alloy. *J. Appl. Phys.*, 35:444, 1964.

[91] M. Muschol, D. Liu, and H. Z. Cummins. Surface-tension-anisotropy measurements of succinonitrile and pivalic acid: Comparison with microscopic solvability theory. *Phys. Rev. A*, 46:1038, 1992.

[92] R. E. Napolitano, Shan Liu, and R. Trivedi. Experimental measurement of anisotropy in crystal-melt interfacial energy. *Interface Science*, 10(12):217–232, 2002.

[93] Patrick Oswald, John Bechhoefer, and Albert Libchaber. Instabilities of a moving nematic-isotropic interface. *Phys. Rev. Lett.*, 58(22):2318–2322, 1987.

[94] S. Pietrowicz, A. Four, S. Jones, S. Canfer, and B. Baudouy. Thermal conductivity and kapitza resistance of cyanate ester epoxy mix and tri-functional epoxy electrical insulations at superfluid helium temperature. *Cryogenics*, 52:100–104, 2012.

[95] A. Pocheau and M. Georgelin. Shape of growth cells in directional solidification. *Phys. Rev. E*, 73:011604, 2006.

[96] William H. Press. *Numerical recipes in C : the art of scientific computing*. Cambridge Univ. Press, Cambridge, 2. edition, 2002.

[97] Antonio Jose Melendez Ramirez. *Experimental investigation of free dendritic growth of succinonitrile-acetone alloys*. PhD thesis, University of Iowa, December 2009.

Bibliography

[98] E. Rolley, S. Balibar, and F. Gallet. The first roughening transition of ^{3}he crystals. *Europhys. Lett.*, 2:247–255, 1986.

[99] Etienne Rolley, Sébastien Balibar, and Francios Graner. Growth shape of ^{3}He needle crystals. *Phys. Rev. E*, 49(2):1500–1506, 1994.

[100] F. Rondelez, W. Urbach, and H. Hervet. Origin of thermal conductivity anisotropy of liquid crystalline phases. *Phys. Rev. Lett.*, 41(15):1058–1062, 1978.

[101] E. R. Rubinstein and M. E. Glicksman. Dendritic grown kinetics and structure. *J. Crys. Growth*, 112:84–86, 1991.

[102] E. R. Rubinstein and M. E. Glicksman. Dendritic grown kinetics and structure. *J. Crys. Growth*, 112:97–110, 1991.

[103] Y. Saito, G. Goldbeck-Wood, and H. Müller-Krumbhaar. Dendritic Crystallization: Numerical Study of the One-Sided Model. *Phys. Rev. Lett.*, 58(15):1541–1543, 1987.

[104] D. A. Saville and P. J. Beaghton. Growth of needle-shaped crystals in the presence of convection. *Phys. Rev. A*, 37(9):3423–3430, 1988.

[105] Y. Sawada, B. Perrin, P. Tabeling, and P. Bouissou. Oscillatory growth of dendritic tips in a three-dimensional. *Phys. Rev. A*, 43(10):5537–5541, 1991.

[106] R. J. Schaefer and S. R. Coriell. Convection-induced distortion of a solid-liquid interface. *Metall. Trans. A*, 15A:2109–2115, 1984.

[107] Dean S. Schrage. A simplified model of dendritic growth in the presence of natural convection. *J. Crys. Growth*, 205:410–426, 1999.

[108] R.F. Sekerka, S.R. Coriell, and G.B. McFadden. Stagnant film model of the effect of natural convection on the dendrite operating state. *J. Crys. Growth*, 154:370–376, 1994.

[109] Robert F. Sekerka. Theory of crystal growth morphology. In G. Müller, J.-J. Métois, and P. Rudolph, editors, *Crystal Growth - from Fundamentals to Technology*, pages 55–93. Elsevier B. V., 2004.

[110] Boris I. Shraiman. Velocity selection and the Saffman-Taylor problem. *Phys. Rev. Lett.*, 56(19):2028–2031, 1986.

[111] Adam J. Simon and Albert Libchaber. Moving interface: The stability tongue and phenomena within. *Phys. Rev. A*, 41(12):7090–7093, 1990.

[112] H. M. Singer and J. H. Bilgram. Three-dimensional reconstruction of xenon dendrites. *Europhys. Lett.*, 68(2):240–246, 2004.

[113] H. M. Singer and J. H. Bilgram. Three-dimensional reconstruction of experimentally grown xenon crystals and characterization of their morphological transitions. *J. Crys. Growth*, 275:e243–e247, 2005.

[114] N. B. Singh and M. E. Glicksman. Determination of the mean solid-liquid interface energy of pivalic acid. *J. Crys. Growth*, 98:573–580, 1989.

[115] N. B. Singh and M. E. Glicksman. Physical properties of ultra-pure pivalic acid. *Thermochim. Acta*, 159:93–99, 1990.

[116] I. Stalder and J. H. Bilgram. Morphology of structures in diffusional growth in three dimensions. *Europhys. Lett.*, 56(6):829–835, 2001.

[117] S. Tanveer. Analytic theory for the selection of a two-dimensional needle crystal at arbitrary Péclet number. *Phys. Rev. A*, 40(8):4756–4769, 1989.

[118] Shakeel H. Tirmizi and William N. Gill. Effect of natural convection on growth velocity and morphology of dendritic ice crystals. *J. Crys. Growth*, 85:488–502, 1987.

[119] X. Tong, C. Beckermann, A. Karma, and Q. Li. Phase-field simulations of dendritic crystal growth in a forced flow. *Phys. Rev. E*, 63:061601, 2001.

[120] Robert Tönhardt and Gustav Amberg. Simulation of natural convection effects on succinonitrile crystals. *Phys. Rev. E*, 62(1):828–836, 2000.

[121] T. Tóth-Katona, T. Börzsönyi, Z. Váradi, J. Szabon, Á. Buka, R. González-Cinca, L. Ramírez-Piscina, J. Casademunt, and A. Hernández-Machado. Pattern formation during mesophase growth in a homologous series. *Phys. Rev. E*, 54(2):1574–1583, 1996.

[122] Martin von Kurnatowski, Thomas Grillenbeck, and Klaus Kassner. Selection theory of free dendritic growth in a potential flow. *Phys. Rev. E*, 87:042405, 2013.

[123] Martin von Kurnatowski and Klaus Kassner. Scaling laws of free dendritic growth in a forced oseen flow. *J. Phys. A*, 47:325202, 2014.

[124] Martin von Kurnatowski and Klaus Kassner. Selection theory of dendritic growth with anisotropic diffusion. *Advances in Condensed Matter Physics*, 2015:529036, 2015.

[125] E. Zauderer. Asymptotic decomposition of partial differential equations. *SIAM J. Appl. Math.*, 35(3):575–592, 1978.

I want morebooks!

Buy your books fast and straightforward online - at one of the world's fastest growing online book stores! Environmentally sound due to Print-on-Demand technologies.

Buy your books online at
www.get-morebooks.com

Kaufen Sie Ihre Bücher schnell und unkompliziert online – auf einer der am schnellsten wachsenden Buchhandelsplattformen weltweit!
Dank Print-On-Demand umwelt- und ressourcenschonend produziert.

Bücher schneller online kaufen
www.morebooks.de

OmniScriptum Marketing DEU GmbH
Heinrich-Böcking-Str. 6-8
D - 66121 Saarbrücken
Telefax: +49 681 93 81 567-9

info@omniscriptum.com
www.omniscriptum.com

Printed by Books on Demand GmbH, Norderstedt / Germany